From the engraving by Jacobus Houbraken

WILLIAM HARVEY

BORN APRIL 1, 1578–DIED JUNE 3, 1657

THE HARVEY LECTURES

DELIVERED UNDER THE AUSPICES OF

The HARVEY SOCIETY of NEW YORK

1996–1997

BY

JUDAH FOLKMAN
DAVID V. GOEDDEL
BRIGID L.M. HOGAN
LEROY HOOD
RICHARD D. KLAUSNER
RICHARD A. LERNER
PAUL NURSE

SERIES 92

1998

WILEY-LISS

A JOHN WILEY & SONS, INC., PUBLICATION

New York • Chichester • Weinheim • Brisbane • Singapore • Toronto

Address All Inquiries to the Publisher
Wiley-Liss, Inc., 605 Third Avenue, New York, NY 10158-0012

This book is printed on acid-free paper. ♾

Published simultaneously in Canada.

ISSN 0073-0874

ISBN 0-471-28326-6

Printed in the United States of America.

10 9 8 7 6 5 4 3 2 1

CONTENTS

HARVEY LECTURES 1996–1997

THE HARVEY SOCIETY*

A SOCIETY FOR THE DIFFUSION OF KNOWLEDGE OF THE MEDICAL SCIENCES

CONSTITUTION

I

This Society shall be named the Harvey Society.

II

The object of this Society shall be the diffusion of scientific knowledge in selected chapters in anatomy, physiology, pathology, bacteriology, pharmacology, and physiological and pathological chemistry, through the medium of public lectures by men and women who are workers in the subjects presented.

III

The members of the Society shall constitute two classes: Active and Honorary members. Active members shall be workers in the medical or biological sciences, residing in the metropolitan New York area, who have personally contributed to the advancement of these sciences. Active members who leave New York to reside elsewhere may retain their membership. Honorary members shall be those who have delivered lectures before the Society and who are not Active members. Honorary members shall not be eligible to office, nor shall they be entitled to a vote.

Active members shall be elected by ballot. They shall be nominated to the Executive Committee and the names of the nominees shall accompany the notice of the meeting at which the vote for their election will be taken.

IV

The management of the Society shall be vested in an Executive Committee to consist of a President, a Vice-President, a Secretary, a Treasurer, and three

*The Constitution is reprinted here for historical interest only; its essential features have been included in the Articles of Incorporation and By-Laws.

other members, these officers to be elected by ballot at each annual meeting of the Society to serve one year.

V

The Annual Meeting of the Society shall be held at a stated date in January of each year at a time and place to be determined by the Executive Committee. Special meetings may be held at such times and places as the Executive Committee may determine. At all meetings ten members shall constitute a quorum.

VI

Changes in the Constitution may be made at any meeting of the Society by a majority vote of those present after previous notification to the members in writing.

THE HARVEY SOCIETY, INC.

A SOCIETY FOR THE DIFFUSION OF KNOWLEDGE OF THE MEDICAL SCIENCES

BY-LAWS

ARTICLE I

Name and Purposes of the Society

SECTION 1. The name of the Society as recorded in the Constitution at the time of its founding in 1905 was the Harvey Society. In 1955, it was incorporated in the State of New York as The Harvey Society, Inc.

SECTION 2. The purposes for which this Society is formed are those set forth in its original Constitution and modified in its Certificate of Incorporation as from time to time amended. The purposes of the Society shall be to foster the diffusion of scientific knowledge in selected chapters of the biological sciences and related areas of knowledge through the medium of public delivery and printed publication of lectures by men and women who are workers in the subjects presented, and to promote the develeopment of these sciences.

It is not organized for pecuniary profit, and no part of the net earnings, contributions, or other corporate funds of the Society shall inure to the benefit of any private member or individual, and no substantial part of its activities shall be carrying on propaganda, or otherwise attempting, to influence legislation.

ARTICLE II

Offices of the Society

SECTION 1. The main office and place of business of the Society shall be in the City and County of New York. The Board of Directors may designate additional offices.

ARTICLE III

Members

SECTION 1. The members of the Society shall consist of the incorporators, members of the hitherto unincorporated Harvey Society, and persons elected from time to time. The members of the Society shall constitute two classes: Active and Honorary Members. Active members shall be individuals with either the Ph.D. or the M.D. degree or its equivalent, residing or carrying on a major part of their work in the New York metropolitan area at the time of their election, who are personally making original contributions to the literature of the medical or biological sciences. Honorary members shall be those who have delivered a lecture before the Society and who are not Active members. Honorary members shall be exempted from the payment of dues. Active members who have remained in good standing for 35 years or who have reached the age of 65 and have remained in good standing for 25 years shall be designated Life members. They shall retain all the privileges of their class of membership without further payment of dues. Honorary members shall not be eligible to office, nor shall they be entitled to participate by voting in the affairs of the society. Volumes of The Harvey Lectures will be circulated only to Active members. Life members will be offered the opportunity to purchase The Harvey Lectures at the cost of the volume. Honorary members will receive only the volume containing their lecture. New Active members shall be nominated in writing to the Board of Directors by an Active member and seconded by another Active member. They shall be elected at the Annual Meeting of the Society by a vote of the majority of the Active members present at the meeting. Members who leave New York to reside elsewhere may retain their membership. Active members who have given a Harvey Lecture and who have moved out of the New York metropolitan area may, if they wish, become Honorary members. Membership in the Society shall terminate on the death, resignation, or removal of the member.

SECTION 2. Members may be suspended or expelled from the Society by the vote of a majority of the members present at any meeting of members at which a quorum is present, for refusing or failing to comply with the By-Laws, or for other good and sufficient cause.

SECTION 3. Members may resign from the Society by written declaration, which shall take effect upon the filing thereof with the Secretary.

ARTICLE IV

Meetings of the Members of the Society

SECTION 1. The Society shall hold its annual meeting of Active members for the election of officers and directors, and for the transaction of such other business as may come before the meeting in the month of January or

February in each year, at a place within the City of New York, and on a date and at an hour to be specified in the notice of such meeting.

SECTION 2. Special meetings of members shall be called by the Secretary upon the request of the President or Vice-President or of the Board of Directors, or on written request of twenty-five of the Active members.

SECTION 3. Notice of all meetings of Active members shall be mailed or delivered personally to each member not less than ten nor more than sixty days before the meeting. Like notice shall be given with respect to lectures.

SECTION 4. At all meetings of Active members of the Society ten Active members, present in person, shall constitute a quorum, but less than a quorum shall have power to adjourn from time to time until a quorum be present.

ARTICLE V

Board of Directors

SECTION 1. The number of directors constituting The Board of Directors shall be seven: the President, the Vice-President, the Secretary, and the Treasurer of the Society, and the four members of the Council. The number of directors may be increased or reduced by amendments of the By-Laws as hereinafter provided, within the maximum and minimum numbers fixed in the Certificate of Incorporation or any amendment thereto.

SECTION 2. The Board of Directors shall hold an annual meeting shortly before the annual meeting of the Society.

Special meetings of the Board of Directors shall be called at any time by the Secretary upon the request of the President or Vice-President or of one-fourth of the directors then in office.

SECTION 3. Notice of all regular annual meetings of the Board shall be given to each director at least seven days before the meeting and notice of special meetings, at least one day before. Meetings may be held at any place within the City of New York designated in the notice of the meeting.

SECTION 4. The Board of Directors shall have the immediate charge, management, and control of the activities and affairs of the Society, and it shall have full power, in the intervals between the annual meetings of the Active members, to do any and all things in relation to the affairs of the Society.

SECTION 5. Council members shall be elected by the members of the Society at the Annual Meeting. One Council member is elected each year to serve for three years, there being three Council members at all times. Vacancies occurring on the Council for any cause may be filled for the unexpired term by the majority vote of the directors present at any meeting at which a quorum is present. Only Active members of the Society shall be eligible for membership on the Council.

SECTION 6. A majority of the Board as from time to time constituted shall be necessary to constitute a quorum, but less than a quorum shall have power to adjourn from time to time until a quorum be present.

SECTION 7. The Board shall have power to appoint individual or corporate trustees and their successors of any or all of the property of the Society, and to confer upon them such of the powers, duties, or obligations of the directors in relation to the care, custody, or management of such property as may be deemed advisable.

SECTION 8. The directors shall present at the Annual Meeting a report, verified by the President and Treasurer, or by a majority of the directors, showing the whole amount of real and personal property owned by the Society, where located, and where and how invested, the amount and nature of the property acquired during the year immediately preceding the date of the report and the manner of the acquisition; the amount applied, appropriated, or expended during the year immediately preceding such date, and the purposes, objects, or persons to or for which such applications, appropriations, or expenditures have been made; and the names of the persons who have been admitted to membership in the Society during such year, which report shall be filed with the records of the Society and an abstract thereof entered in the minutes of the proceedings of the Annual Meeting.

ARTICLE VI

Committees

SECTION 1. The Board of Directors may appoint from time to time such committees as it deems advisable, and each such committee shall exercise such powers and perform such duties as may be conferred upon it by the Board of Directors subject to its continuing direction and control.

ARTICLE VII

Officers

SECTION 1. The officers of the Society shall consist of a President, a Vice-President, a Secretary, and a Treasurer, and such other officers as the Board of Directors may from time to time determine. All of the officers of the Society shall be members of the Board of Directors.

SECTION 2. The President shall be the chief executive officer of the Society and shall be in charge of the direction of its affairs, acting with the advice of the Board of Directors. The other officers of the Society shall have the powers and perform the duties that usually pertain to their respective offices, or as may from time to time be prescribed by the Board of Directors.

SECTION 3. The officers and the directors shall not receive, directly or indirectly, any salary or other compensation from the Society, unless authorized by the concurring vote of two-thirds of all the directors.

SECTION 4. The officers shall be elected at the Annual Meeting of the Active members. All officers shall hold office until the next Annual Meeting and until their successors are elected or until removed by vote of a majority of the directors. Vacancies occurring among the officers for any cause may be filled for the unexpired term by the majority vote of the directors present at any meeting at which a quorum is present. Officers must be Active members of the Society.

ARTICLE VIII

Fiscal Year—Seal

SECTION 1. The fiscal year of the Society shall be the calendar year.

SECTION 2. The seal of the Society shall be circular in form and shall bear the words "The Harvey Society, Inc., New York, New York, Corporate Seal."

ARTICLE IX

Amendments

SECTION 1. These By-Laws may be added to, amended, or repealed, in whole or in part, by the Active members or by the Board of Directors, in each case by a majority vote at any meeting at which a quorum is present, provided that notice of the proposed addition, amendment, or repeal has been given to each member or director, as the case may be, in the notice of such meeting.

OFFICERS OF THE HARVEY SOCIETY

1996–1997

STUART AARONSON, President
LENNART PHILIPSON, Vice President
SCOTT EMMONS, Treasurer
PAUL B. LAZAROW, Secretary

COUNCIL

1996–1997

EVA B. CRAMER
LORRAINE J. GUDAS
DAVID I. HIRSH
MARSHALL HOROWITZ
KENNETH MARIANS
PETER MODEL

CORPORATE SPONSORS

Amgen Inc.
Boehringer Ingelheim Pharmaceuticals, Inc.
Bristol Myers-Squibb Co.
Hoffman-LaRoche Pharmaceuticals
Pfizer Inc.

PREFACE:
A BRIEF HISTORY OF THE HARVEY SOCIETY, NEW YORK

The Harvey Society was founded April 1, 1905 by 13 New York scientists and physicians, who met at 9 East 74th Street, the home of physiologist Graham Lusk. The stated purpose of the Society was to forge a "closer relationship between the purely practical side of medicine and the results of laboratory investigation." The group included Samuel J. Meltzer, William H. Park, Edward K. Dunham, James Ewing, Frederick S. Lee, Christian Herter, Simon Flexner, George B. Wallace, Theodore C. Janeway, Phoebus A. Levene, and Eugene L. Opie. The meeting was also attended by John J. Abel, a noted pharmacologist from The Johns Hopkins University School of Medicine.

Founding officers:

President	Graham Lusk
Vice-President	Simon Flexner
Secretary	George B. Wallace
Treasurer	Frederick S. Lee

The Harvey Society lectures reflect "the evolution of physiology and physiological chemistry into biochemistry and the development of molecular biology from the roots of bacteriology and biochemistry" in this century. One of the important facets of this lecture series is that they are published annually. The Harvey Society is continuing to achieve its primary goal of providing "a series of distinguished lectures in the life sciences" and, in fact, "there are few lectures of such distinction delivered uninterruptedly over a period of more than 70 years in any other city in the world."[1]

Harvey once wrote, "My trust is in my love of truth and the candour of cultivated minds." For as long as those who are invited to deliver a Harvey Lecture heed these words, they honour the name of William Harvey, and the Harvey Society of New York has a high and noble future.[1]

---

[1]Exerpted from an article in *Perspectives in Biology and Medicine,* Summer 1978, pages 524–535, by A.G. Bearn and D.G. James.

MAKING ENZYMES

RICHARD A. LERNER, CARLOS F. BARBAS III,
AND KIM D. JANDA

The Scripps Research Institute and the Skaggs Institute for Chemical Biology, Departments of Chemistry and Molecular Biology, La Jolla, California

I. INTRODUCTION

IN this paper we discuss the science of antibody catalysis that takes as its goal the production of protein enzymes in real time. By way of example, we build two rather complex enzymes and in so doing take a tour through the main intellectual and technical aspects of the field. But first one may ask why one would wish to build new enzymes since some 4000 or so enzymes already exist in nature. A simple answer might be that even this large number of enzymes cannot encompass all the chemical transformations that one wishes to catalyze. Thus, there are many important reactions for which there are no enzymes, and even when enzymes exist, their substrate restrictions may preclude their use. But, there is a more compelling reason that cannot be better put than the words inscribed by the late Richard Feynman on his blackboard at Cal. Tech., where he wrote, "What I cannot create I do not understand" (Fig. 1). This is extremely important in the study of high works of evolution such as enzymes, where the contributors to rate acceleration act in concert so that each is perturbed by the other, making it very difficult to isolate individual components of catalysis. However, as one makes an enzyme in real time, it becomes possible to learn much about the physical organic chemistry of catalysis and ultimately the ability of protein-binding pockets to do chemistry. Implied in these ideas are issues of how proteins impart strain, overcome entropic barriers, deal with highly reactive intermediates, control stereoelectronic parameters, and use their hydrophobic interiors to accomplish chemistry that would be impossible in solution or even in the gas phase. How then does one begin to make an antibody catalyst?

The Harvey Lectures, Series 92, pages 1–40

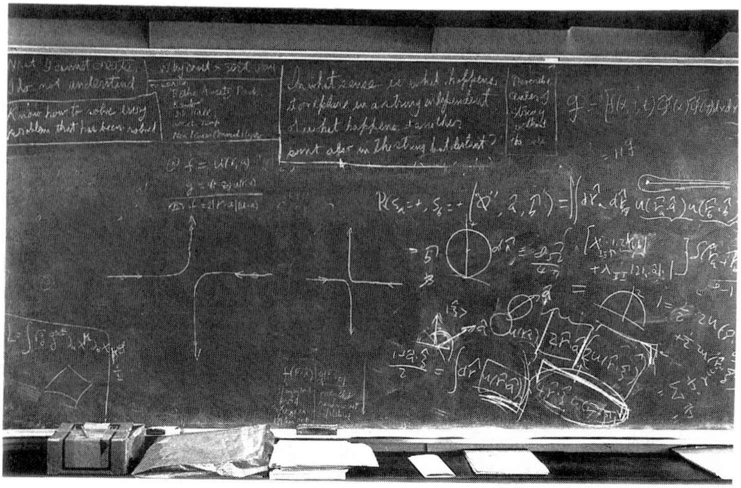

Fig. 1. Blackboard of Richard Feynman from the Cal. Tech. archives.

Much of what we need to know is illustrated on a blackboard photographed at the end of Paulings Silliman lecture in 1948 (Fig. 2). Pauling was discussing what was known and what remained to be done. When he turned to enzymes and antibodies, he said that they differ in that antibodies bind to ground states, whereas enzymes bind to transition states, thereby lowering the energy barrier and increasing the rate of the reaction. Thus, on this blackboard we have all the ingredients for the science of antibody catalysis. First, we have a fundamental theory of catalysis that suggests what kinds of compounds should induce antibodies that catalyze reactions. Second, a protein diversity system exists that is capable of induction by the experimenter and will bind to anything that be synthesized.

However, before we go into this, we need to understand, from a chemical point of view, what kind of machine the antibody molecule is because it, after all, will become the main instrument of our catalysis, and success will depend on its ability to do chemistry. In the early days of antibody catalysis much was said about features of the antibody molecule that would preclude its ability to do chemistry. Some said it was too rigid, whereas others said it was too flexible. Some felt that the pockets were too deep, while others felt that antibodies were incapable of generating pockets of sufficient depth to

Fig. 2. Blackboard from the end of Linus Pauling's Silliman lecture.

exclude solvent. The answer to these issues should have been obvious to anyone who thought deeply about the system from a chemical point of view. As a combinatorial system that is selected based on binding energy, the molecule is neither too deep nor too shallow nor too flexible nor too rigid, but is rather anything that is necessary to achieve binding energy. Thus, if binding energy gains can be made by constructing a deep pocket to exclude solvent, this happens, whereas if a collection of small binding interactions at approximated surfaces is favorable to binding, then this happens.

So what kind of chemical machine is the antibody molecule? First, it is in every sense of the word the product of combinatorial chemistry, starting with its synthesis, which will not be discussed here. It is naive, which is a feature often overlooked but extremely important to a system that must evolve binding in real time, because it is much easier to shape a malleable protein than to dismantle a "hard wired" binding pocket. The system achieves the all-important linkage of recognition and replication that is, of course, the main feature of many biological systems, including sperm and egg interaction, development and differentiation, and antibody induction. The linkage of recognition and replication in immune cells allows for the special process of somatic mutation to yield antibodies with higher and higher affinity for the inducing antigen. If recognition and replication were not linked in immune cells, there would not be any means for propagation of the successful mutations that ultimately lead to high binding energy.

Thus, the immune system is a biological system that depends on the presence of an inducer for replication while at the same time producing a product (the antibody molecule) that removes the inducer. If one throws in somatic mutation and the linkage of recognition and replication, then one has a system that by its intrinsic nature drives toward the highest binding energy. The systems conbinatorics (Fig. 3) is the product of the random interaction of two chains, each approximating three essentially random peptide loops of variable length in three-dimensional space. The system, thus, gives a target size to chemists who would build diversity systems for recognition. If one can cause six chains, each containing about ten building blocks selected from a pool of twenty functionalities, to be approximated in three-dimen-

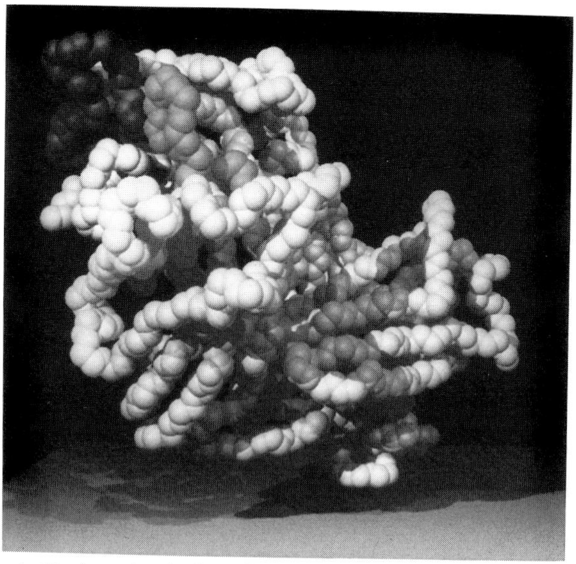

Fig. 3. Antibody molecule from the work of David Davies showing six CDRs (top left) that form the peptide combinatorial when they are approximated in three-dimensional space. Rendering by Professor Arthur Olson from coordinates provided by Davies and colleagues (PNAS, **74,** 5140–5144, 1977).

sional space, a huge number of molecules will be selectively recognized by the diverse members of such a library of molecules.

The chemical output of the antibody diversity is illustrated in Figures 4 and 5, in which the simple issue of shape complementarily is explored. We see the solution to six different binding interactions when the system was induced with different molecules varying from another protein to the rigid small organic molecule fluorescein (Averalo et al., 1993a,b, 1994; Gibson et al., 1988; Rini et al., 1992; Sheriff et al., 1987; Smith-Gill et al., 1987; Stanfield et al., 1990). When confronted with the protein lysozyme, the antibody molecule evolves to approximate the greatest surface possible, thereby maximizing the potential for productive interaction. When confronted with a small organic hapten, the antibody evolves to bury it deeply, thereby achieving the binding energy implicit in the hydrophobic effect, and when con-

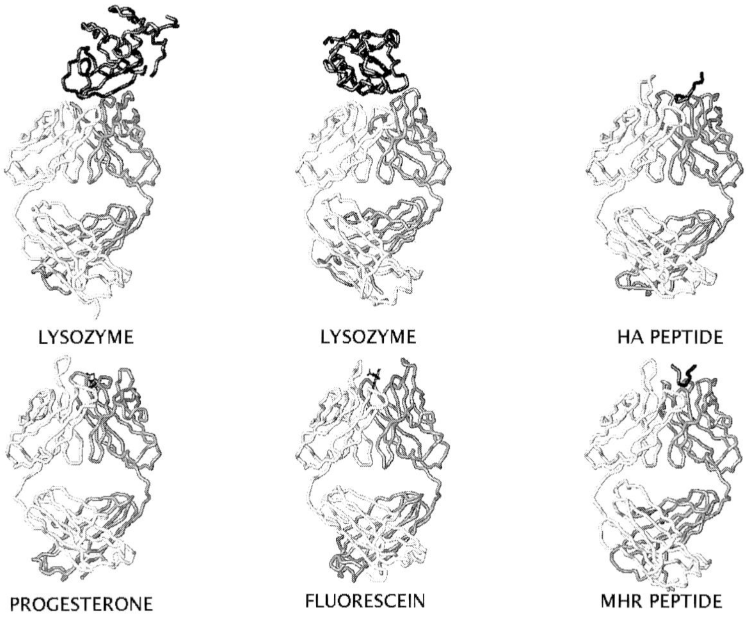

Fig. 4. Antibody crystal structures of six different molecules, with the antibody structure on the bottom and the complexed antigen on the top. Notice how the diversity system selects for an antibody of the proper shape. Figure courtesy of Prof. Ian Wilson.

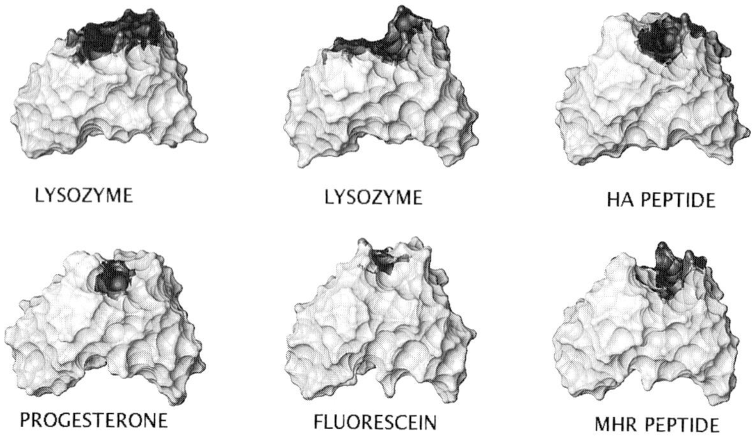

Fig. 5. Same as Figure 4, but now the antigen is removed and the contact surface of the antibody is highlighted to emphasize the excellent molding of the antibody to the nature of the antigen. Figure courtesy of Prof. Ian Wilson.

fronted with a helical peptide that is intermediate between a protein and a small organic molecule, the antibody gives an intermediate solution and builds a moderately deep groove. Thus, the system is a "great physical chemist," and if it makes a move you do not understand, you had better think again about how it could lead to increased binding energy.

However, there is a fundamental problem in harnessing this system for catalysis in that the system selects for binding by any means, whereas in catalysis the exact way in which binding occurs is the key. Therefore, if one revisits a lysozyme anti-lysozyme complex (Amit et al., 1986) and strips away all of the molecule except those amino-acid residues in the antibody that contact the antigen, one sees a wonderful array of functionalities that could be put in the service of catalysis (Figs. 6, 7). There are primary amines, alcohols, phenols carboxylates, imidazoles, indoles, and guanidiniums that are even more powerful because they exist in a hydrophobic pocket. The problem, of course, is that they are organized for binding and not for the highly specific interaction of the functionalities with the transition state that is the essence of catalysis. Thus, to achieve catalysis the experimenter must cause these functionalities to be arranged in highly specific ways around the transition state of a reaction. This is done, as it were, by giving a chemical

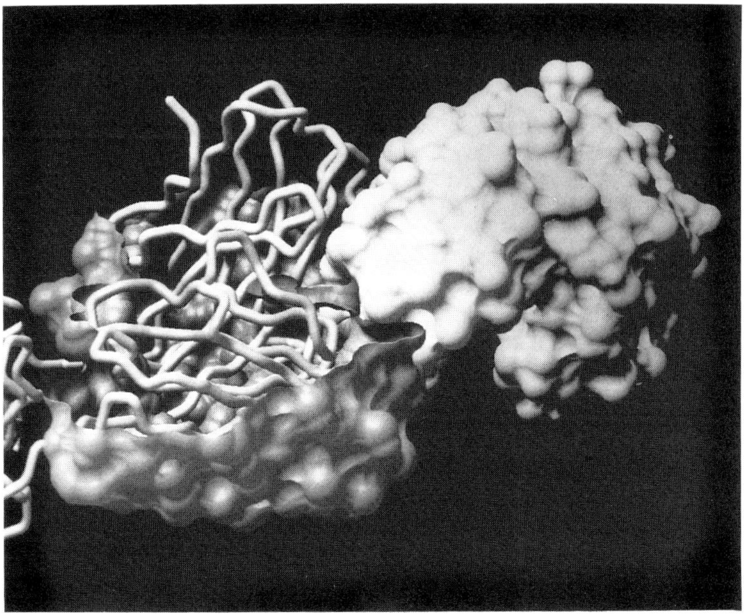

Fig. 6. Structure by Poljak and colleagues of lysozyme complexed with Anti-lysozyme, showing excellent approximation of interacting surfaces. Rendering by Professor Arthur Olson from coordinates of Poljak et al. (*Science,* **233,** 747–753).

education to the protein via the nature of the inducer or hapten that the chemist makes. Often the hapten is an analogue of the transition state of the reaction. Figure 8 can be used to illuminate the philosophy of the various approaches that chemists use to make catalysts. Everyone would like to achieve catalytic perfection, but each desires to achieve it in a different way, depending on the field one is studying.

If one wishes to study evolution, then the highest form of the art is to enter at zero (or close to zero) and let the system run toward perfection. Only random selection is allowed, and instruction by the experimenter is not fair. Here, the art is in how the system is built so as to be able to take advantage of the powerful processes of mutation and selection. Antibody catalysis is, on the other hand, a chemical science where the highest art form is to give detailed chemical instructions to the induction system so as to enter as close to catalytic perfection as possible. Thus, the science is largely

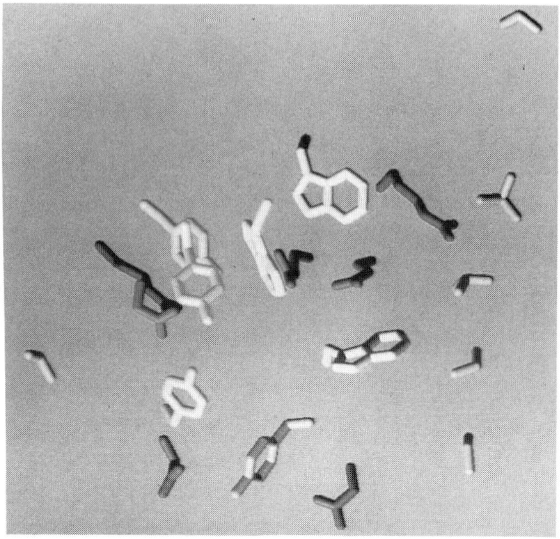

Fig. 7. Same structure as in Figure 6, but now only interacting amino acid side chains of antibody are showing, emphasizing diverse chemical entities that contact antigen.

instructional, and the better the instructions the better the catalyst. This does not, however, mean that the system is unable to teach us about evolution, because we have learned much about the way protein molecules react to chemical instructions, and the way that instructions are received ultimately dictates the concert of complexities that make up an enzyme. Indeed, during studies of antibody catalysis we have seen convergent evolution, induced fit, covalent catalysis, and unusual reaction outcomes such as cationic cyclopropanation (*Isr. J. Chem.*, 1996).

Fig. 8. A scale for evolution of catalysis.

In the initial demonstrations of antibody catalysis, simple reactions such as ester hydrolysis were chosen because the reaction mechanism was understood and there was a large difference in shape and charge between the transition state and either the substrate or the products (Fig. 9). Thus, during hydrolysis the planar uncharged ester proceeds along the reaction coordinate toward the charged tetrahedral intermediate, which collapses to the acid and alcohol products. The phosphonate molecule is an analogue of this high-energy intermediate because it is tetrahedral and negatively charged, and the phosphate–oxygen bond is approximately 20% longer than the carbon-oxygen bond, thereby mimicking the stretching of the latter bond in the transition state of the reaction after attack of water or hydroxide ion at the electrophilic carbonyl carbon. Such phosphonates were, and continue to be, very successful in inducing esterolytic antibodies, and as the crystal structures of esterolytic catalytic continue to emerge, we are learning how faithfully the antibody molecule accepts the instruction of the chemist to build a pocket that accepts the tetrahedral shape and neutralizes the developing negative charge.

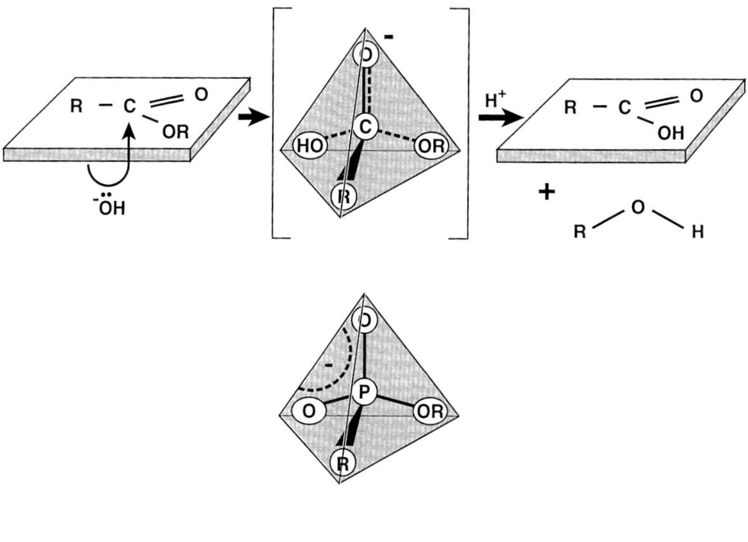

Fig. 9. Scheme illustrating some of the early highlights of the first antibody catalysts made that hydrolyzed esters using phosphonates as transition state analogues.

Some of the ability of the antibody molecule to take instructions pertinent to catalysis can be seen in the first structure of a catalytic antibody for the Claisen reaction solved by Hilvert and Wilson and their colleagues (Haynes et al., 1994, 1996; Hilvert et al., 1988) (Figs. 10, 11). The Claisen rearrangement is formally a 3,3 sigmatrophic rearrangement in which the enol pyruvate side chain must move from its extended conformation to lie over the cyclohexadienal system in the transition state (Fig. 10). It was already known that the oxabicyclic system designed to mimic this state of affairs was an inhibitor of the natural enzyme, and when the Schultz group (Jackson et al., 1988) and the Hilvert group (Haynes et al., 1996) independently made antibodies to this transition state analogue, they were able to obtain catalytic antibodies that, in the case of the Schultz group, were highly efficient relative to the natural enzyme. When the crystal structure of the antibody-complexed to the hapten was obtained, it illustrated how well the antibody binding pocket adapted to the instructions.

The overall geometry of the pocket and its asymmetry tell us much about why this antibody is a catalyst. The pocket deeply buries the oxabicyclic hapten and presumably the substrate sequestering the reaction from water,

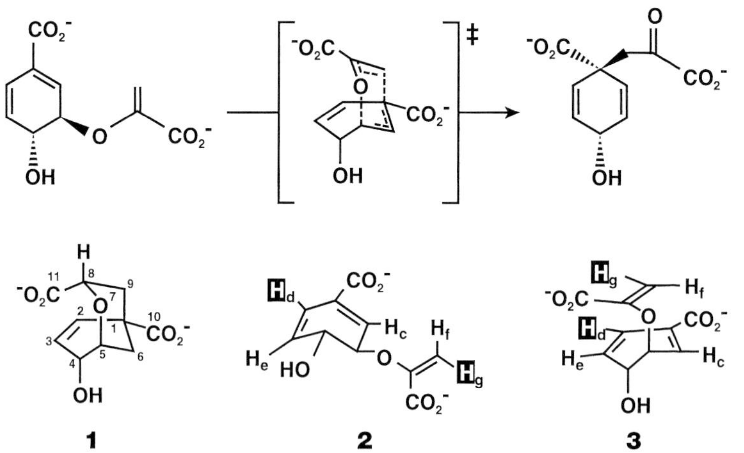

Fig. 10. Reaction path from Hilvert *et al.* of the Claisen rearrangement showing transition state and transition state analogue (compound 1), extended conformation (structure 2), and reactive conformation (structure 3). The bold hydrogen atoms show the approximation that was detected by NMR.

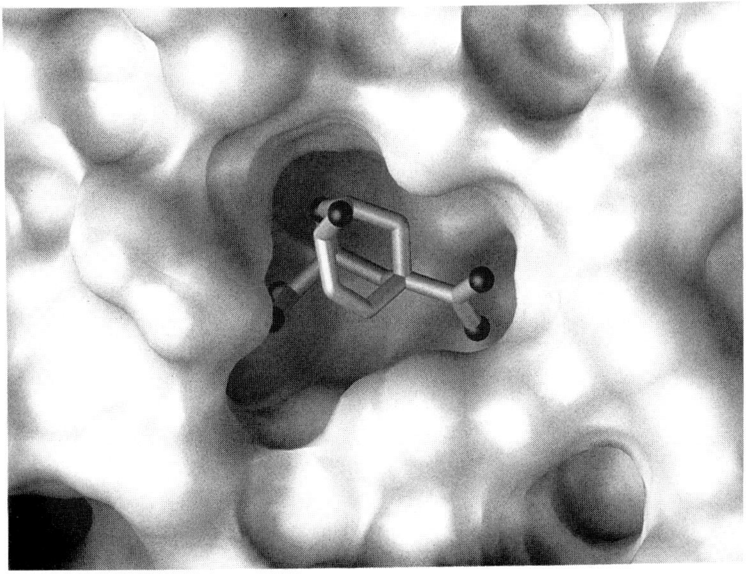

Fig. 11. Structure from Wilson, Hilvert, and their colleagues of Claisen catalytic antibody with transition state analogue complexed, illustrating excellent acceptance of the chemists's instructions by the antibody. Note that the antibody does not accept the extended conformation of the substrate, thus forcing the enol pyruvate side chain to lie over the cyclohexadienal system. Also, the asymmetry of the antibody binding pocket explains the enantioselectivity of the catalysis.

which might not be so important in the chemical details of this largely concerted transformation but clearly must help in the entropic contribution to the binding energy. The shape of the binding pocket will not allow binding of the substrate in its extended conformation, and thus should force the enol pyruvate side chain over the cyclohexadienal system, which is entropically favorable for the transformation. Finally, the asymmetry of the binding pocket explains the stereochemical outcome of the reaction in that only one antipode of the substrate is accepted by the antibody. Thus, the antibody faithfully accepts the instructions of the chemist and in so doing becomes a catalyst. This is not to say that the instructions were perfect, particularly since there may be considerable enolate character to the transition state of this reaction and, thus, introduction of some charge into the antigen might be desirable.

The general ideas discussed above have led to the induction of many catalysts that are discussed in the some 650 publications. Rather than compiling these reactions, we illustrate in Table I some of the milestones (*Isr. J. Chem.*, references therein, 1996). This list is not meant to be exhaustive or even inclusive, but rather indicates papers that taught major lessons or indicated directions that the field might take.

II. Teaching the Advanced Course to the Immune System

In the early experiments in antibody catalysis, scientists were satisfied with induction of antibodies that dealt with gross chemical details of the transition state such as overall shape and charge and, for bimolecular reactions, approximation of reactants. These were features that were easy to deal with, and it is clear how one might achieve complementarity in the antibody binding pocket. The next advance came as scientists, emboldened by initial success, devised methods to introduce countercharges in a regioselective way, thereby inducing and directing more complex reactions such as those that require lowering the energy of highly reactive carbocations that appear along the reaction coordinate. However, it was felt that the exact mechanism by which the antibody achieved catalysis was somewhat up to chance since there was no way to dictate exactly how the antibody molecule would achieve binding energy.

Thus, let us, by way of example, consider nucleophilic catalysis. This is one of the most powerful solutions that nature has evolved for catalysis of difficult transformations such as amide bond hydrolysis, where the initial adduct is an acyl enzyme. This mechanism requires a nucleophile in the active site that is of appropriate pKa to attack the carbonyl of the scissile bond. Covalent catalysis requires extreme organization of the binding pocket in that it requires precise positioning in three-dimensional space of an amino acid side chain (or concert of side chains as in the serine proteases) relative to atoms in the substrate. Given these requirements, it was not easy to see how one might call forward these mechanistic details in an ordinary immunological induction, and many thought that these details were best left to improvements that might be achieved by site-specific mutagenesis or, in the worst case, would forever remain as the sole purview of nature's highly evolved enzymes. However, the concept and practice of reactive immunization has changed all this and is allowing chemists to induce antibodies that catalyze reactions according to defined mechanisms, including the otherwise-elusive covalent catalysis (Wirsching et al., 1995; Wagner et al., 1995).

TABLE I. MILESTONES IN THE INDUCTION OF CATALYSTS

1947:	**Linus Pauling proposes the principle of transition-state stabilization to explain enzyme catalysis.** Pauling, L. (1948). *Am. Sci.* **36,** 51–58.
1969:	**William P. Jencks proposes the idea of catalytic antibodies.** Jencks, W.P. (1969). "Catalysis in Chemistry and Enzymology," McGraw-Hill, New York.
1986:	**Antibody catalysis via transition-state stabilization is first demonstrated.** (a) Tramontano, A., Janda, K.D., and Lerner, R.A. (1986). *Science* **234,** 1566–1570. (b) Pollack, S.J., Jacobs, J.W., and Schultz, P.G. (1986). *Science* **234,** 1570–1573.
1987:	**Enantioselectivity in antibody-catalyzed reaction is achieved.** Napper, A.D., Benkovic, S.J., Tramontano, A., and Lerner, R.A. (1987). *Science* **237,** 1041–1043.
1988:	**Antibody catalyzes amide hydrolysis.** Janda, K.D., Schloeder, D., Benkovic, S.J., and Lerner, R.A. (1988). *Science* **241,** 1188–1191.
	Antibody catalyzes a pericyclic reaction: the Claisen rearrangement. (a) Hilvert, D., and Nared, K.D. (1988). *J. Am. Chem. Soc.* **110,** 5593–5594. (b) Hilvert, D., Carpenter, S.H., Nared, K.D., and Auditor, M.-T.M. (1988). *Proc. Natl. Acad. Sci. U.S.A.* **85,** 4953–4955. (c) Jackson, D.Y., Jacobs, J.W., Sugasawara, R. Reich, S.H., Bartlett, P.A., and Schultz, P.G. (1988). *J. Am. Chem. Soc.* **110,** 4841–4842.
1989:	**Antibody catalyzes a pericyclic reaction: the Diels-Alder reaction.** (a) Hilvert, D., Hill, K.W., Nared, K.D., and Auditor, M.-T.M. (1989). *J. Am. Chem. Soc.* **111,** 9261–9262. (b) Braisted, A.C., and Schultz, P.G. (1990). *J. Am. Chem. Soc.* **112,** 7430–7431.
	Strategies for eliciting general bases in the active site are developed. Shokat, K.M., Leumann, C.J., Sugasawara, R., and Schultz, P.G. (1989). *Nature* **338,** 269–271.
	Antibodies are expressed and produced in trangenic plants. Hiatt, A.C., Cafferkey, R., and Bowdish, K. (1989). *Nature* **342,** 76–78.
1990:	**Substrate distortion leads to ferrocatalase activity.** Cochran, A.G., and Schultz, P.G. (1990). *Science* **249,** 781–783.
	Antibodies catalyze oxidations with prophyrin cofactors. (a) Cochran, A.G., and Schultz, P.G. (1990). *J. Am. Chem. Soc.* **112,** 9414–9415. (b) Keinan, E., Sinha, S.C., Sinha-Bagehi, A., Benory, E., Gozi, M., Eshhar, Z., and Green, B.S. (1990). *Pure Appl. Chem.* **62,** 2013–2019.
1991:	**Single-chain catalytic antibody is constructed and characterized.** Gibbs, R.A., Posner, B.A., Filpula, D.R., Dodd, S.W., Finkelman, M.A.J., Lee, T.K., Wroble, M., Whitlow, M. and Benkovic, S.J. (1991). *Proc. Natl. Acad. Sci. U.S.A.* **88,** 4001–4004.

(continues)

TABLE I. MILESTONES IN THE INDUCTION OF CATALYSTS *(continued)*

	Transesterification catalysis using ping-pong and induced-fit mechanisms is observed. (a) Wirsching, P., Ashley, J.A., Benkovic, S.J., Janda, K.D., and Lerner, R.A. (1991). *Science* **252,** 680–685. (b) Jacobson, J.R., Prudent, J.R., Kochersperger, L., Yonkovich, S., and Schultz, P.G. (1992). *Science* **256,** 365–367.
	Antibody catalyzes multistep *cis-trans* isomerization reaction. Jackson, D.Y., and Schultz, P.G. (1991). *J. Am. Chem. Soc.* **113,** 2319–2321.
	Esterolytic antibodies resolve each of four diastereomeric esters. Kitazume, T., Lin, J.T., Yamamoto, T., and Yamazaki, T.J. (1991). *J. Am. Chem. Soc.* **113,** 8573–8575.
	The phage-display technology for screening catalytic Fab fragments is introduced. Barbas, C.F., III, Kang, A.S., Lerner, R.A., and Benkovic, S.J. (1991). *Proc. Natl. Acad. Sci. U.S.A.* **88,** 7978–7982.
	***In vivo* catalysis is demonstrated with antibodies elicited by genetic selection.** Tang, Y., Hicks, J.B., and Hilvert, D. (1991). *Proc. Natl. Acad. Sci. U.S.A.* **88,** 8784–8786.
1992:	**Highly enantioselective proton transfer is catalyzed by an antibody.** Reymond, J.-L., Janda, K.D., and Lerner, R.A. (1992). *J. Am. Chem. Soc.* **114,** 2257–2258.
	Antibody catalysis involves an unnatural cofactor: ketone reduction. (a) Nakayama, G.R., and Schultz, P.G. (1992). *J. Am. Chem. Soc.* **114,** 780–000. (b) Suh, J., Oh, E., Kim, S.H., Lee, C.S., and Jeong, G. (1991). *Bull. Korean Chem. Soc.* **12,** 352–354.
	Naturally occurring catalytic antibodies that hydrolyze DNA are discovered. Shuster, A., Gololobov, G.V., Kvashuk, O.A., Bogomolova, A.E., Smirnov, I.V., and Gabibov, A.G. (1992). *Science* **256,** 665–667.
1993:	**Antibody catalyzes a disfavored reaction: violation of the "Baldwin rules."** Janda, K.D., Shevlin, C.G., and Lerner, R.A. (1993). *Science* **259,** 490–493.
	Antibody catalyzes a disfavored reaction: exo-Diels-Alder reaction. Gouverneur, V.E., Houk, K.N., de Pascual-Teresa, B., Beno, B., Janda, K.D., and Lerner, R.A. (1993). *Science* **262,** 204–208.
	Antibody catalyzes a disfavored reaction: regio/enantioselective ketone reduction. Hsieh, L.C., Yonkovich, S., Kochersperger, L., and Schultz, P.G. (1993). *Science* **260,** 337–339.
	CatELISA—a facile method to screen catalytic activity. Tawfik, D.S., Green, B.S., Chap, R., Sela, M., and Eshhar, Z. (1993). *Proc. Natl. Acad. Sci. U.S.A.* **90,** 373–377.
	Catalytic antibodies show therapeutic applications: degradation of cocaine. Landry, D.W., Zhao, K., Yang, G.X., Glickman, M., and Georgiadis, T.M. (1993). *Science* **259,** 1899–1901.

(continues)

TABLE I. MILESTONES IN THE INDUCTION OF CATALYSTS *(continued)*

	Catalytic antibodies show therapeutic application: prodrug activation. (a) Miyashita, H., Karaki, Y., Kikuchi, M., and Fujii, I. (1993). *Proc. Natl. Acad. Sci. U.S.A.* **90,** 5337–5340. (b) Campbell, D.A., Gong, B., Kochersperger, L.M., Yonkuovich, S., Gallop, M.A., and Schultz, P.G. (1994). *J. Am. Chem. Soc.* **116,** 2165–2166.
1994:	**Antibody catalyzes a non-enzymic reaction: asymmetric epoxidation of unfuctionalized olefins.** Koch, A., Reymond, J.-L., and Lerner, R.A. (1994). *J. Am. Chem. Soc.* **116,** 803–804.
	Antibody catalyzes a non-enzymic reaction: oxy-Cope rearrangement. Braisted, A.C., and Schultz, P.G. (1994). *J. Am. Chem. Soc.* **116,** 2211–2212.
	Antibody catalyzes a non-enzymic reaction: oxy-generation with periodate. Hsieh, L.C., Stephans, J.C., and Schultz, P.G. (1994). *J. Am. Chem. Soc.* **116,** 2167–2168.
	Remarkable control of chemical reactivity is achieved: syn elimination to *cis* olefin. Cravatt, B.F., Ashley, J.A., Janda, K.D., Boger, D.L., and Lerner, R.A. (1994). *J. Am. Chem. Soc.* **116,** 6013–6014.
	Remarkable control of chemical reactivity is achieved: stereoselective oxime formation. Uno, T., Gong, B., and Schultz, P.G. (1994). *J. Am. Chem. Soc.* **116,** 1145–1146.
	Remarkable control of chemical reactivity is achieved: cationic cyclization. Li, T., Janda, K.D., Ashley, J.A., and Lerner, R.A. (1994). *Science* **264,** 1289–1293.
	Enantioselective multigram scale synthesis with a catalytic antibody is achieved. Reymond, J.-L., Reber, J.-L., and Lerner, R.A. (1994). *Angew. Chem. Int. Ed. Engl.* **33,** 475–477.
	Antibody catalyzes formation of a peptide bond. (a) Jacobsen, J.R., and Schultz, P.G. (1994). *Proc. Natl. Acad. Sci. U.S.A.* **91,** 5888–5892. (b) Hirschmann, R., Smith, A.B. III, Taylor, C.M., Benkovic, P.A., Taylor, S.D., Yager, K.M., Sprengeler, P.A., and Benkovic, S.J. (1994). *Science* **265,** 234–237.
	Crystal structure of a catalytic antibody shows similarity to a serine protease active site. Zhou, G.W., Guo, J., Huang, W., Fletterick, R.J., and Scanlan, T.S. (1994). *Science* **265,** 1059–1064.
	Catalytic antibodies are elicited via heterologous immunization. (a) Suga, H., Ersoy, O., Williams, S.F., Tsumuraya, T., Margolis, M.N., Sinsky, A.J., and Masamune, S. (1994). *J. Am. Chem. Soc.* **116,** 8432. (b) Tsumuraya, T., Suga, H., Meguro, S., Tsunakawa, A., and Masamune, S. (1995). *J. Am. Chem. Soc.* **117,** 11390–11396.
	Structure-activity relationships of a catalytic antibody show similarity to its natural counterpart. Haynes, M.R., Stura, E.A., Hilvert, D., and Wilson, I.A. (1994). *Science* **263,** 646–652.

(continues)

TABLE I. MILESTONES IN THE INDUCTION OF CATALYSTS *(continued)*

Esterolytic antibodies deprotect acylated carbohydrates regio/stereoselectivity. Iwabuchi, Y., Miyashita, H., Tanimura, R., Kinoshita, K., Kikuchi, M., and Fujii, I. (1994). *J. Am. Chem. Soc.* **116,** 771–772.

1995: **Reactive species is sequestered from water by a catalytic antibody: ketalization in aqueous media.** Shabat, D., Itzhaky, H., Reymond, J.-L., and Keinan, E. (1995). *Nature* **374,** 143–146.

The first total synthesis of a natural product via antibody catalysis is achieved. Sinha, S.C., and Keinan, E. (1995). *J. Am. Chem. Soc.* **117,** 3653–3654.

The reactive immunization strategy is introduced. Wirsching, P., Ashley, J.A., Lo, C.-H.L., Janda, K.D., and Lerner, R.A. (1995). *Science* **270,** 1775–1782.

Efficient aldolase catalytic antibodies use the enamine mechanisms of natural enzymes. Wagner, J., Lerner, R.A., and Barbas, C.F. III. (1995). *Science* **270,** 1797–1800.

Crystal structure shows mechanistic convergence in esterase-like catalytic antibodies. Charbonnier, J.B., Carpenter, E., Gigant, B., Golinelli-Pimpaneau, B., and Eshhar, Z. (1995). *Proc. Natl. Acad. Sci. U.S.A.* **92,** 11721–11725.

1996: **Antibodies perturb the coordingaiton sphere of a transition metal complex.** Ghosh, P., Shabat, D., Kumar, S., Sinha, S.C., Grynszpan, F., Li, J., Noodleman, L., and Keinan, E. (1996). *Nature* **382,** 339–441.

Chemical selection for catalysis in combinational antibody libraries is achieved. Janda, K.D., Lo, L.-C., Lo, C.-H., Sim, M.-M., Wong, C.-H., and Lerner, R.A. (1996). *Science* **275,** 945–948.

Structural studies provide new insight into the immunological evolution of antibody catalysis. Patten, P.A., Gray, N.S., Yang, P.L., Marks, C.B., Wedemayer, G.J., Boniface, J.J., Stevens, R.C., and Schultz, P.G. (1996). *Science* **271,** 1086–1091.

Antibody catalyzes cationic cyclopropanation. Li, T., Janda, K.D., and Lerner, R.A. (1996). *Nature* **379,** 326–327.

Antibodies catalyze reactions of a broad range of substrates while maintaining high reaction sensitivity. Tanaka, E., Kinoshita, K., Tanimura, R., and Fujii, I. (1996). *J. Am. Chem. Soc.* **118,** 2332–2339.

Mutagenesis leads to improved catalytic efficiency. Ulrich, T., et al. *Science* (submitted for publication).

Antibody catalyzes tandem cationic polycyclization. Hasserodt, J., Janda, K.D., and Lerner, R.A. (1996). *J. Am. Chem. Soc.* (submitted for publication).

Before beginning a discussion of reactive immunization, we should discuss the goals of ordinary immunization. Since its inception the process of immunization has dealt with inert antigens. In fact, among immunologists the inertness of an antigen has been a matter of pride. One could not attend many immunochemistry meetings in the early days of anti-peptide antibodies without hearing that "my antigen is more inert (i.e., more native) than yours." The reason for this is clear. Since one usually wishes to induce antibodies to react with native proteins or stable organic compounds, the closer one comes to this state the better. For example, since we believe that viruses are infectious when their proteins are in their native conformation, it makes most sense to duplicate this in a vaccine so that the induced antibodies will neutralize the virus. The process of reactive immunization (Figs. 13, 14, 15) is just the opposite to this in that it attempts to immunize with the most reactive compound possible so as to call forward a chemical reaction in the antibody binding pocket during the process of induction (Wirsching et al., 1995; Wagner et al., 1995). Then, this same chemical reaction is used later as part of the catalytic mechanism. *In a sense one immunizes with a chemical reaction rather than a chemical.*

To illustrate the principle we can consider the induction of antibodies that efficiently hydrolyze ester substrates via a covalent mechanism. The phosphodiester antigen shown in Figs. 13–15 can be expected to induce antibodies that might bind in two fundamentally different ways. In the first instance, the antibody might utilize a series of non-covalent interactions

Concept of Reactive Immunization

1. **Conventional immunogens are inert**
2. **Reactive immunization utilizes reactive immunogens such that the antibody induction process involves a chemical transformation**
3. **That same chemical transformation becomes part of the mechanism of the subsequent catalytic event**

In a sense one immunizes with a chemical reaction

Fig. 12. Major concepts of reactive immunization.

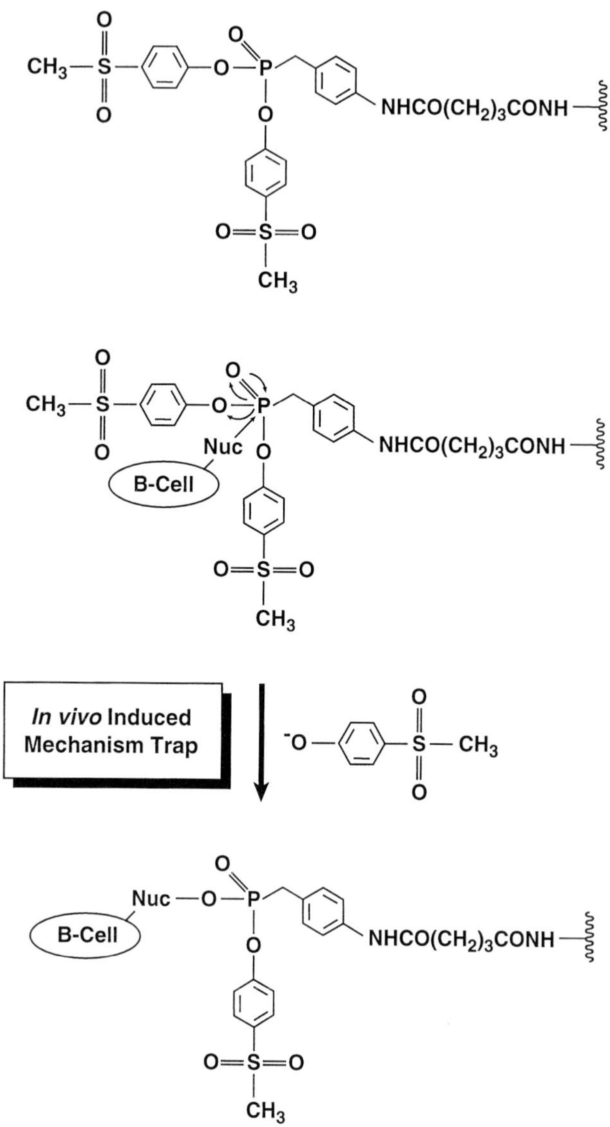
CH3—S—O—P—NHCO(CH2)3CONH
O=S=O
CH3
Nuc
B-Cell
In vivo Induced
Mechanism Trap
⁻O—S—CH3
Nuc—O—P

(*a*)

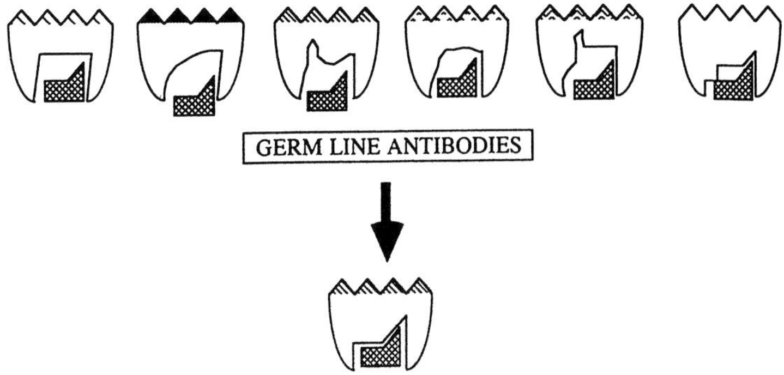

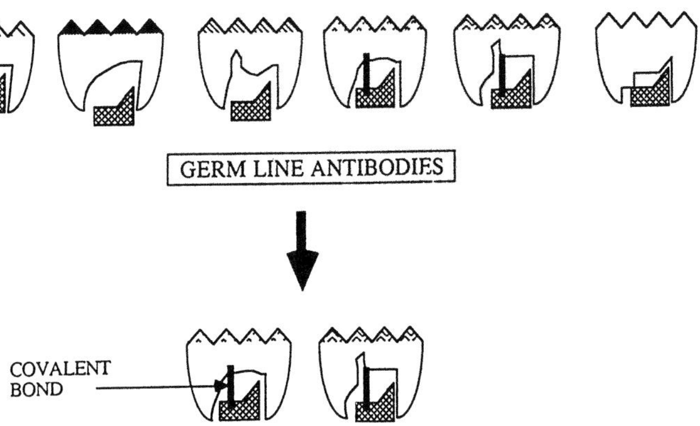

(*b*)

Fig. 13. Scheme by which a reactive immunogen induces an immune response where the resultant antibodies react covalently with the antigen.

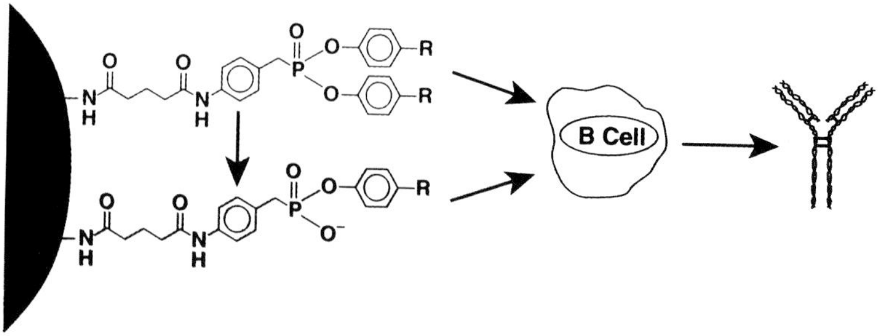

Fig. 14. An actual phosphodiester antigen used to induce antibodies that are efficient esterases. The central idea is to select for antibodies that react with the phosphate atom and cause phenol to leave. Reproduced from *Science* **270,** 1775–1782 (1995).

that in the aggregate ultimately add up to the binding energy achieved. And, whereas all these interactions were not in place in the germline encoded protein, they would be achieved by the process of somatic mutation that, in this case, selects on the basis of superior binding affinity. There is, however, another mode of interaction that calls for the occasional positioning of a reactive nucleophile near the phosphonate atom such that a phosphonylation reaction occurs leading to covalent attachment of the antigen to the antibody binding pocket (Figs. 14, 15). In one single step the antibody has

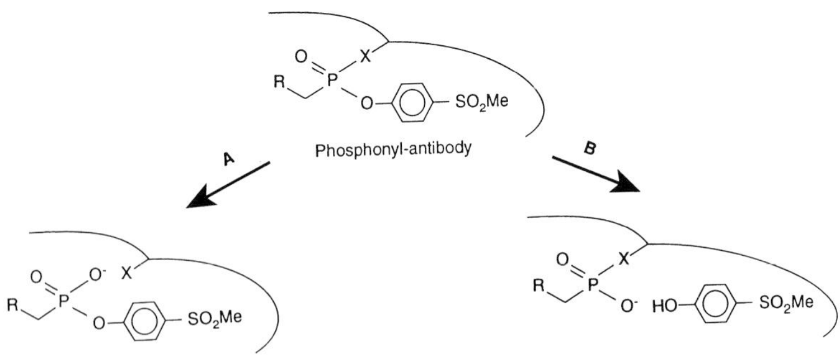

Fig. 15. A scheme showing the chemical opportunities that can take place in the antibody combining pocket when it binds to the phosphodiester antigen. X=the nucleophile in the antibody. Reproduced from *Science* **270,** 1775–1782 (1995).

achieved binding energy far greater than otherwise possible since a covalent bond has been achieved. The competition between clones should cease at this point, and the output phase of antibody production process should turn to production of antibodies capable of covalent interaction with the antigen. Incidently, such a covalent process would be expected to have two interesting consequences beyond that of catalytic efficiency. First, the distance between the resultant antibody and the germline molecule depends on when the reactive nucleophile is generated. In a sense, it is a way of "sequencing" the process of somatic mutation since once the nucleophile appears, the process of somatic mutation should cease to yield improved antibodies. Second, when the nucleophile appears early in the process of somatic mutation, the antibody may be relatively promiscuous since a high degree of refinement of the binding pocket to more closely fit the antigen was no longer necessary to achieve good affinity. This has consequences for catalysis since such antibodies made by reactive immunization might be expected to accept a broad range of substrates.

The key now is to put the chemical reaction that occurred during the process of reactive immunization into service for catalysis of defined mechanism. This is not difficult because one already knows a great deal about the chemical capability of the binding pocket. We already know that a reactive nucleophile is present and is capable of reacting with an electrophile that is at or near the three-dimensional position of the atom in the antigen that trapped it in the first place. All that is left is to build a cognate substrate that is capable of using the chemistry already in place. Thus, the nucleophile that reacted with the phosphate atom of the antigen should be capable of interacting with the carbonyl carbon of amide and ester substrates so long as the carbonyl carbon occupies a position in the antibody pocket at or near that of the phosphate atom of the antigen. This is easily ensured by using an antigen that, other than having a phosphoryl functionality, is the same as or similar to the substrate. In the most general sense, reactive antigens such as those just described are related to the mechanism-based inhibitors used in enzymology. These inhibitors are used to inhibit enzymes when the mechanism is known or probe the mechanism when there are questions. Here, mechanism-based inhibitors are used in the reverse sense, with the central idea being that if one immunizes with a mechanism-based inhibitor, one will induce mechanism.

One of the most remarkable catalytic antibodies ever prepared was obtained by reactive immunization (Wagner et al., 1995). Before describing it, let us introduce what is perhaps the most difficult problem in the generation of new catalysts. This is the need to obtain catalysts that have the efficiency of natural enzymes but are much broader in scope. The reason that this is difficult is that efficiency is accompanied by an ensemble of highly specific binding interactions that necessarily limit scope in terms of substrate choice. Thus, for the enzymes evolved in nature, there is high efficiency coupled with high specificity, and this is all to the good because in metabolism one does not want side reactions. However, for organic synthesis, one wishes to obtain enzymes of broad scope because one does not wish to make a new enzyme every time the substrate changes slightly. Thus, in organic synthesis the most useful enzymes are those that catalyze a class of reactions regardless of the nature of the groups appended to the functional group that is to be modified. As discussed above, there is a loophole in the immune process that can be exploited, which allows one to obtain both efficiency and promiscuity. In the ordinary process of immunization, the process of somatic refinement leads to a series of improved interactions with the antigen that, as with enzymes, leads to higher and higher binding energy with the attendant increase in specificity. However, in reactive immunization, the antigen makes a covalent bond with the antibody binding site, giving the clone that has made this interaction a competitive advantage over all others. Thus, any antibody that makes a covalent bond has better binding energy than all other antibodies, regardless of how many noncovalent improvements they make during the process of somatic refinement. When the functionality that leads to the covalent bond appears early in the process of antibody evolution, further refinement is aborted and that clone becomes dominant in the immune process (Fig. 13b). The result of these events is that one has an antibody in which chemical potential has been installed in a binding pocket that is otherwise not highly refined. This should lead to a pocket that is capable of both efficiency and promiscuity.

Let us examine this process in the construction of antibody aldolases that are both efficient and broad in scope. The aldol addition reaction is one of the most important carbon–carbon bond-forming processes in chemistry and biology. In nature this reaction is accomplished by enzymes that are highly efficient but narrow in scope. In chemistry most of the catalysts in use are stoichiometric and require blocking groups and preformed enolates (Fig. 16). There are two classes of natural enzymes that either use the transition metal zinc to polarize the carbonyl or a highly reactive lysine residue to prepare a carbon nucleophile donor that adds to a second carbonyl carbon via an eneamine mechanism.

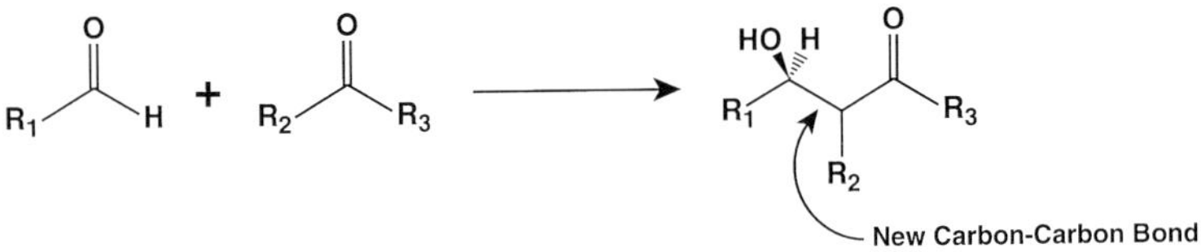

Chemistry:

Preformation of an enolate

Until recently, use of stoichiometric amounts of chiral auxiliary

Protecting groups necessary

Aldolases:

High donor specificity

Fig. 16. Features of the control of the aldol reaction that are used in chemistry and enzymology. Each has its advantages and deficiencies. In antibody catalysis one wishes an amalgamation of the best features of each.

Let us examine the second class of enzymes in detail (Fig. 17). An aldol donor such as acetone enters the enzyme pocket where it encounters a reactive lysine of perturbed pKa to form a hemiaminal that is dehydrated to a protonated Schiff base. The Schiff base lowers the activation energy for deprotonation to form the eneamine that is the nascent carbon nucleophile that adds to an acceptor carbonyl such as an aldehyde. The centerpiece of this chemical mechanism is the eneamine functionality. Given the complexity of this mechanism one might have thought that it would be extremely difficult to install it in an antibody binding pocket and that one would be best advised to attempt to prepare antibody aldolases that use the zinc mechanism. However, the power of reactive immunization is such that even a mechanism as complex as this is easy to install by choosing the right immunogen.

In Figure 18 the reactive immunization process to prepare aldolases is illustrated. The immunogen is a 1,3 dicarbonyl functionality that is chosen to select for a reactive lysine in the immunization process. When this lysine appears during the process of somatic refinement, it can attack one of the carbonyl carbons to form a Schiff base that can tautomerize to an eneamine that, because of the second carbonyl, forms stable vinylogous amide. The vinylogous amide is expected to have a strong absorption outside the range of proteins (*ca.* 316 nM), and thus any antibody that has made this interaction can be detected. Although this selection to first approximation seems simple, it is powerful in terms of chemistry in that it selects not only for a

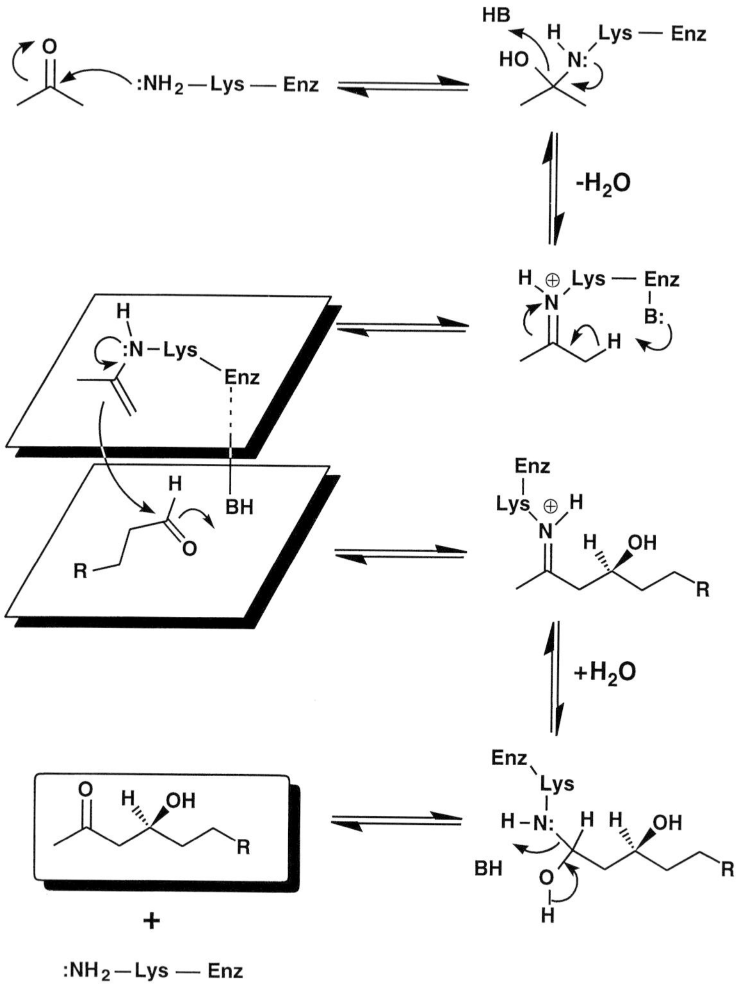

Fig. 17. Mechanistic feature of the natural aldolase enzyme featuring the central role of the eneamine. Reproduced from *Science* **270,** 1797–1800 (1995).

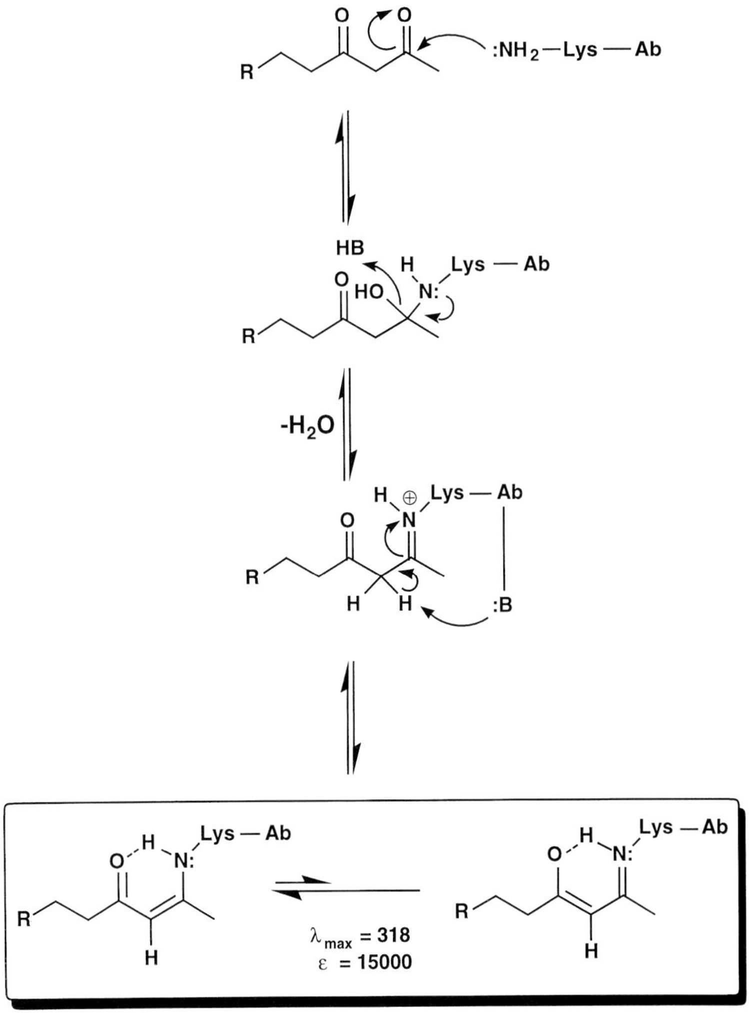

Fig. 18. Reactive immunogen used to select an antibody aldolase that proceeds by an eneamine mechanism.

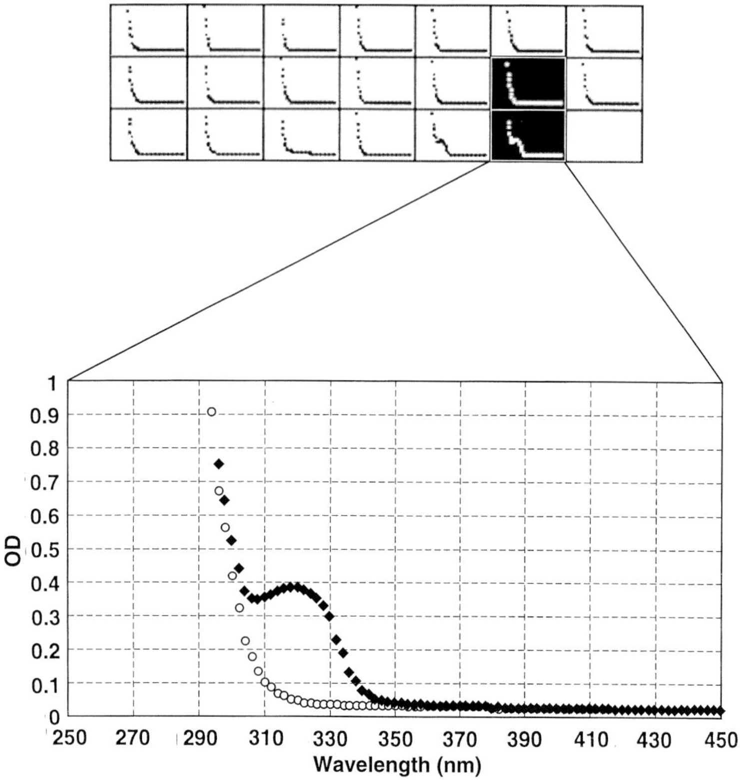

Fig. 19. (A) Screening of antibodies for the formation of vinylogous amide intermediate. Hapten 1,5 was added to 20-μM solutions of each antibody in PBS buffer (pH 7.5) in a microtiter plate format. (B) Antibodies with aldolase catalytic activity presented the typical absorption maximum of the vinylogous amide at 316 nm, whereas none of the inactive antibodies did. Two antibodies out of 20 formed the intermediate and were subsequently determined to be catalytic. Reproduced from *Science* **270,** 1797–1800 (1995).

Fig. 21. The active sites of 38C2 were titrated with acetylacetone. The Ab concentration (20 μM) was fixed. Acetylacetone (0 to 4.5 eq) was added and the absorption measured at 316 nm. The intersection of the two lines corresponds to a ratio of 1.9 of acetylacetone to antibody 38C2, indicating that both binding sites of the antibody form the enamine adduct. Reproduced from *Science* **270,** 1797–1800 (1995).

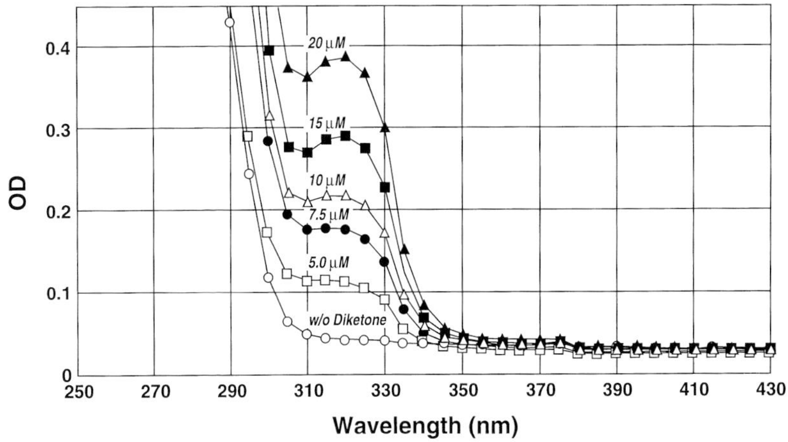

Fig. 20. Determination of the extinction coefficient of a fixed concentration (100 μM) of hapten was added to the indicated concentrations (micromolar) of 33F12. The antibody enamine complex could easily be detected at an antibody concentration as low as 2 μM. Reproduced from *Science* **270,** 1797–1800 (1995).

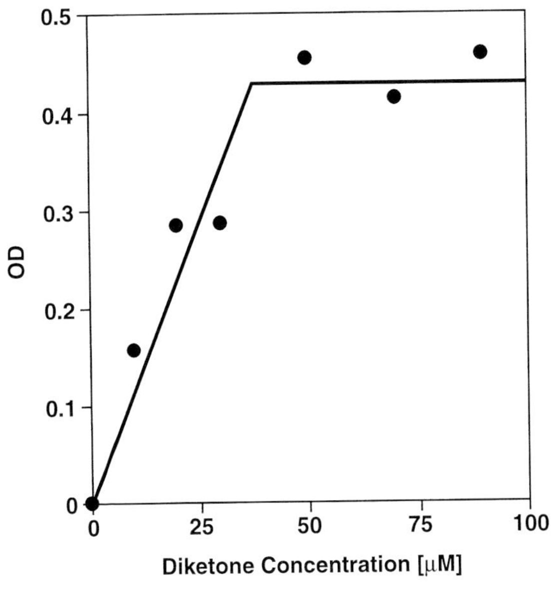

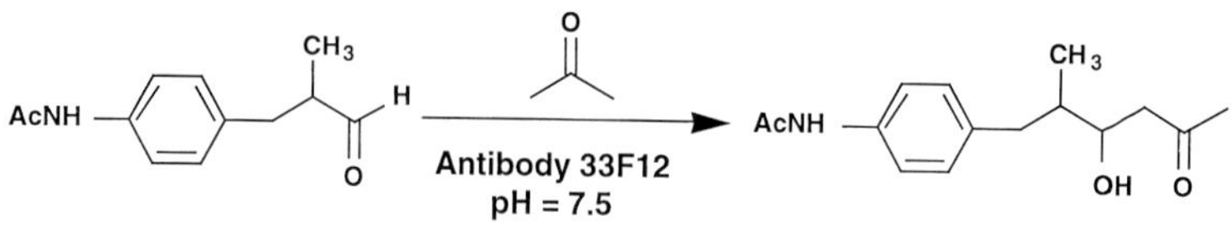

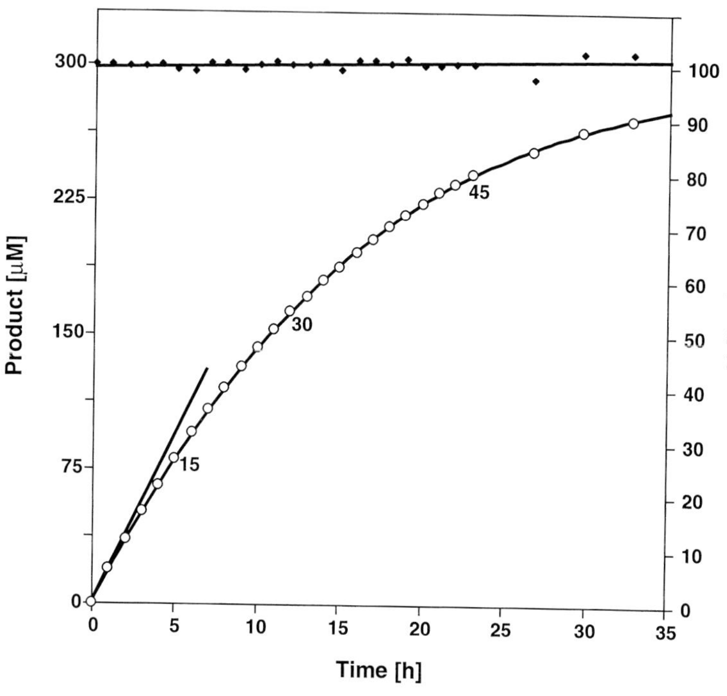

Fig. 22. The aldol addition reaction of aldehyde and acetone was monitored over a 36-hour period in the presence of 1.5% catalyst (20) (Ac, acyl). The catalyst showed multiple turnovers (~ 2 turnovers per hour) and virtually no product inhibition. A 90% conversion could be obtained in the presence of excess acetone (5% v/v) to minimize the retro–aldol reaction. The prefect mass balance (top line) indicates that no side reactions, such as elimination or polymerization, occurred over that period. Thus, the antibody–catalyzed aldol reaction is an exceptionally mild method of C–C bond formation.

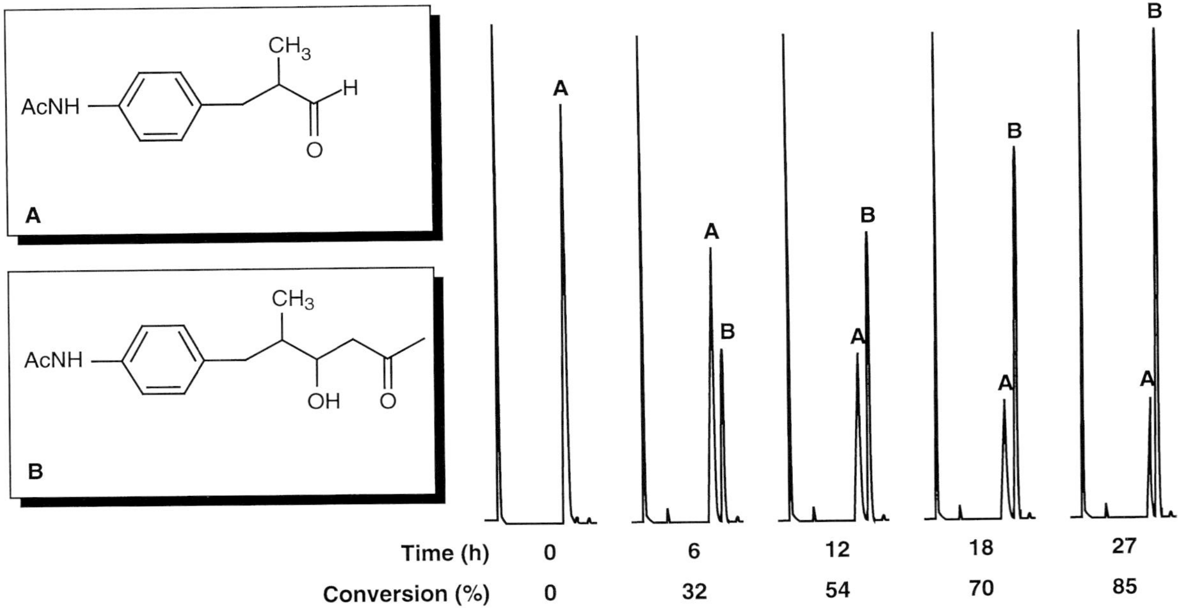

Fig. 23. HPLC confirmation of the mass balance experiment shown in Figure 22. During the reaction one has either starting material (A) or produce (B), but nothing else.

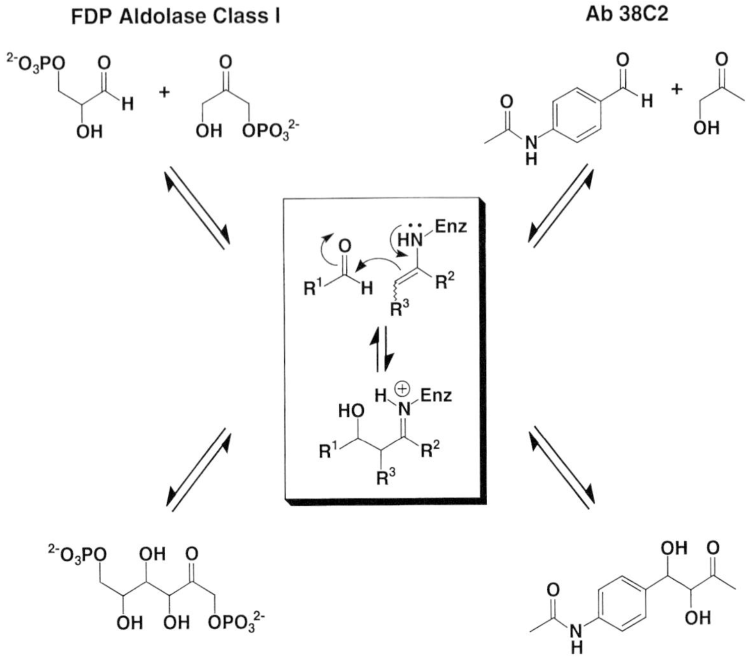

Fig. 24. Comparison of the reaction mechanisms of the natural aldolase enzyme and the antibody aldolase, each using one of their best substrates (adapted from *Science,* **278,** 2085–2092 (1997)).

lysine but that it is also one that is in an environment that both perturbs its reactivity and allows it to approximate the carbonyl carbon it must react with. If one thinks about it, one realizes that this is a powerful selection process along the road to catalysis in that it programs a rather precise chemical environment.

In Figures 3 and 19–21, one sees the recordings from a kinetic plate reader in which a series of hybridomas that resulted from the immunization with the 1,3-dicarbonyl immunogen are studied. We see that two of the antibodies show an additional absorption at 316 nm, as predicted for formation of the vinylogous amide. Although it is beyond the scope of this paper, we have studied the pKa of the reactive lysine in these antibodies and, with Ian Wilson's group (Barbas et al., 1997) have solved the crystal structure of one

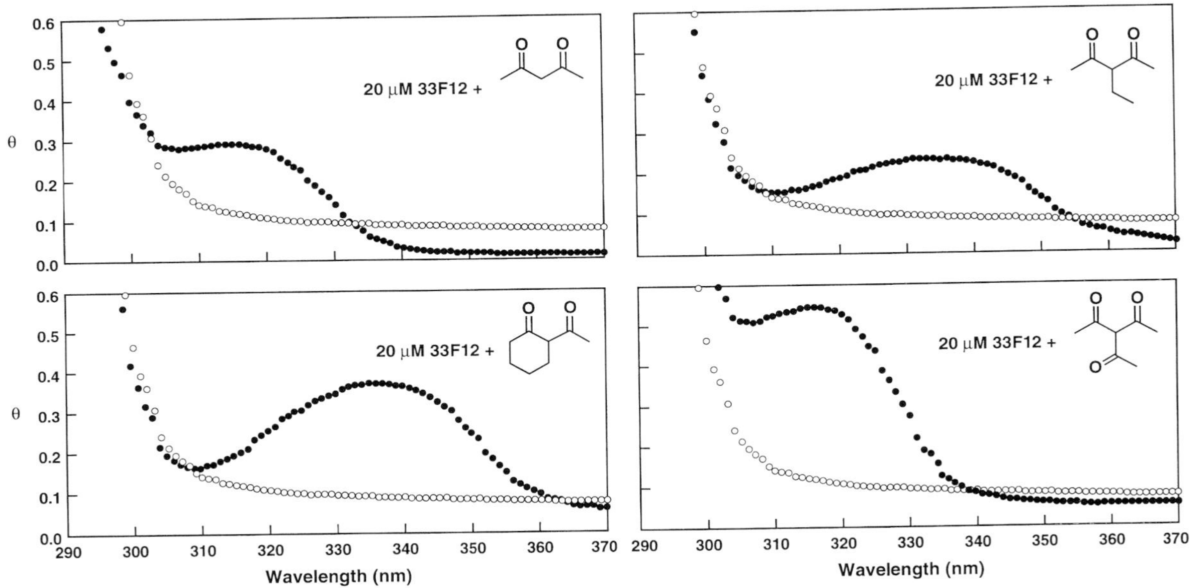

Fig. 25. Spectroscopic signal as antibody accepts a variety of 1,3 diketones to form vinylogous amide. As expected, the maximum absorption shifts slightly according to the substitution pattern of the diketone.

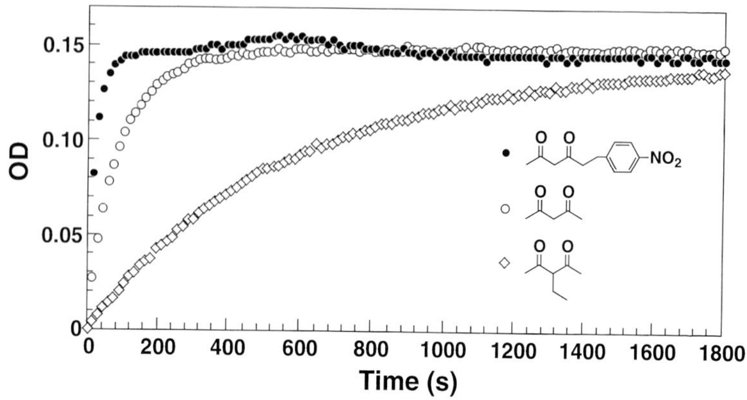

Fig. 26. Rate of formation of Schiff base as a function of different 1,3 dicarbonyl compounds. In all cases the formation of the Schiff base is fast.

of the antibodies. The lysine is deeply buried in the antibody binding pocket and, as with the natural enzymes, has a highly perturbed pKa of 6.5. In the case of the antibody, the perturbed pKa likely results from the fact that it appeared in a highly hydrophobic environment such that protonation is disfavored. The antibody is a highly efficient aldolase (Figs. 22, 23) with a rate enhancement for the crossed aldol reaction that is only one hundred-fold less than nature's own aldolase, which is the centerpiece of sugar metabolism and essential for all life (Fig. 24).

However, what is most remarkable is the broad in scope of the antibody aldolase. Using spectroscopic studies, it was learned early on that this antibody would take a large range of substrates into its binding pocket and form the vinylogous amide (Figs. 25 and 26), but its remarkable range was only to be discovered later (Fig. 27). It accomplishes over 100 different aldol additions and condensations, some of which lead to quite valuable chemical entities such as the Wieland-Miescher ketone formed by the antibody in a

Fig. 28. The catalysis of construction of the Wieland-Miescher ketone by the antibody. The reaction is formally a Robinson annulation in which the antibody catalyzes both steps (Michael addition and aldol cyclodehydration) and controls the stereochemistry.

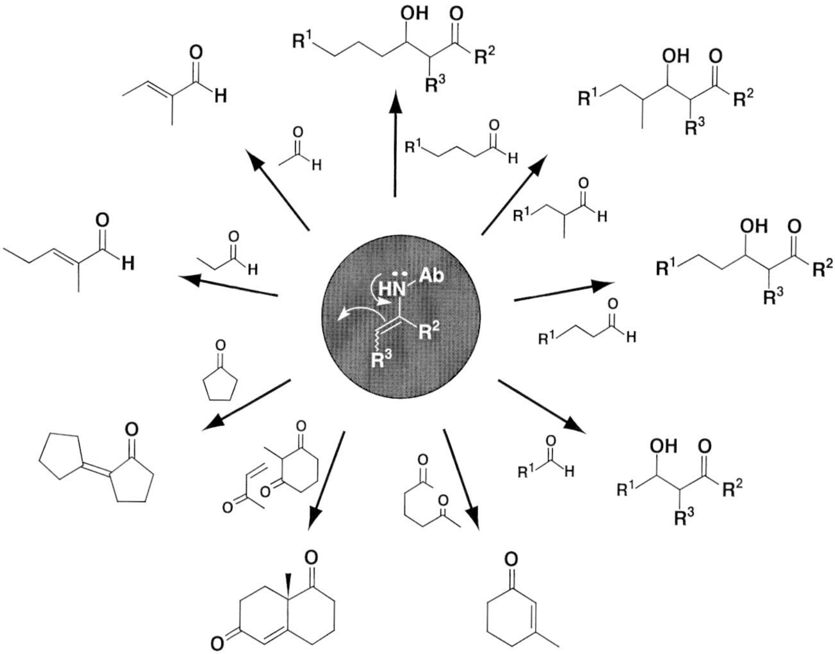

Fig. 27. Some of the aldol additions and condensations of the aldolase catalytic antibody, illustrating its remarkable scope. Reproduced from *Science,* **278,** 2085–2092 (1997).

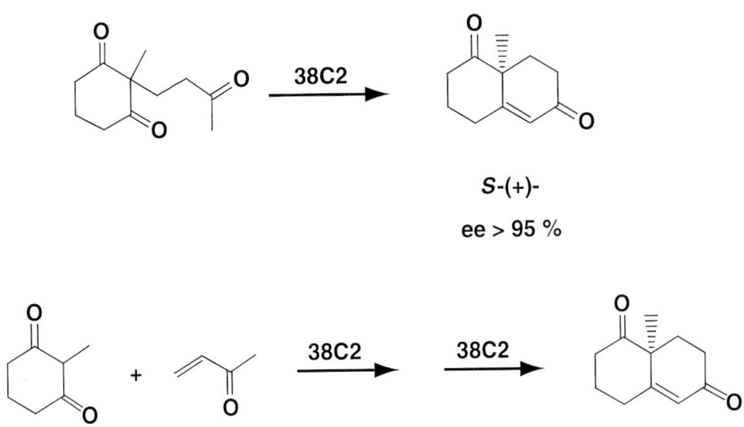

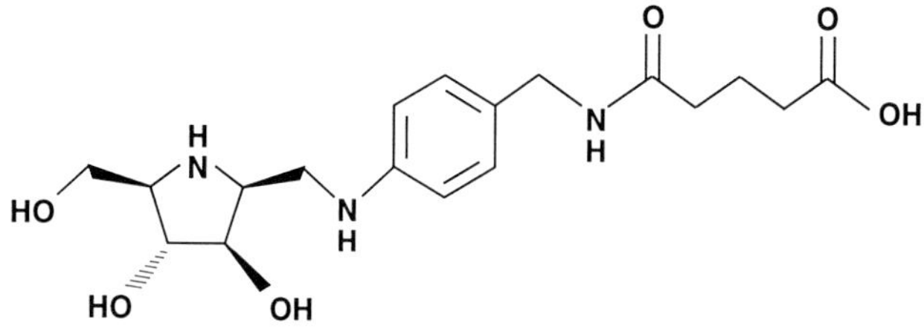

Hapten

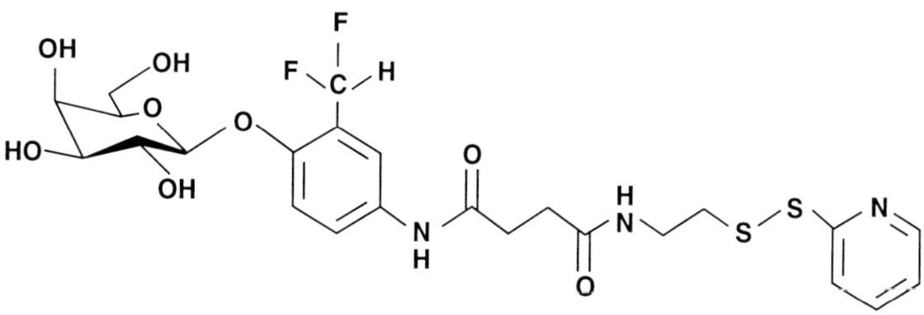

Mechanism Based Panning Reagent

Fig. 29. Structure of a hapten used to induce glycosidase antibodies and a complex used to select a phage that carries out chemistry in the binding pocket of the surface antibody. The hapten is a mimic of the half-chair conformation of the transition state as well as giving a positive charge when the amine is protonated. The positive charge is used to induce a countercharge to stabilize the positively charged oxacarbenium functionality that appears in the transition state of glycosidic bond hydrolysis. The panning reagent has three essential components: (1) A sugar recognition portion; (2) a leaving group that becomes an alkylating agent when the glycosidic bond is cleaved; (3) a disulfide bond that can be disrupted by reduction to liberate phage that is covalently attached to a matrix as a result of the cleavage of the glycosidic bond. Adapted from *Science,* **275,** 945–948 (1997).

Robinson annulation reaction (Fig. 28) (Zhong et al., 1997). Thus, in this case, reactive immunization has led to a catalytic antibody that rivals the natural enzyme but is much broader in scope. These findings would seem to validate the concept that simultaneously solves the reactivity and scope problem.

There is another way of thinking about what happened in the generation of this antibody that is quite interesting in terms of evolution. In the generation of enzymes that are the result of the process of natural selection, the

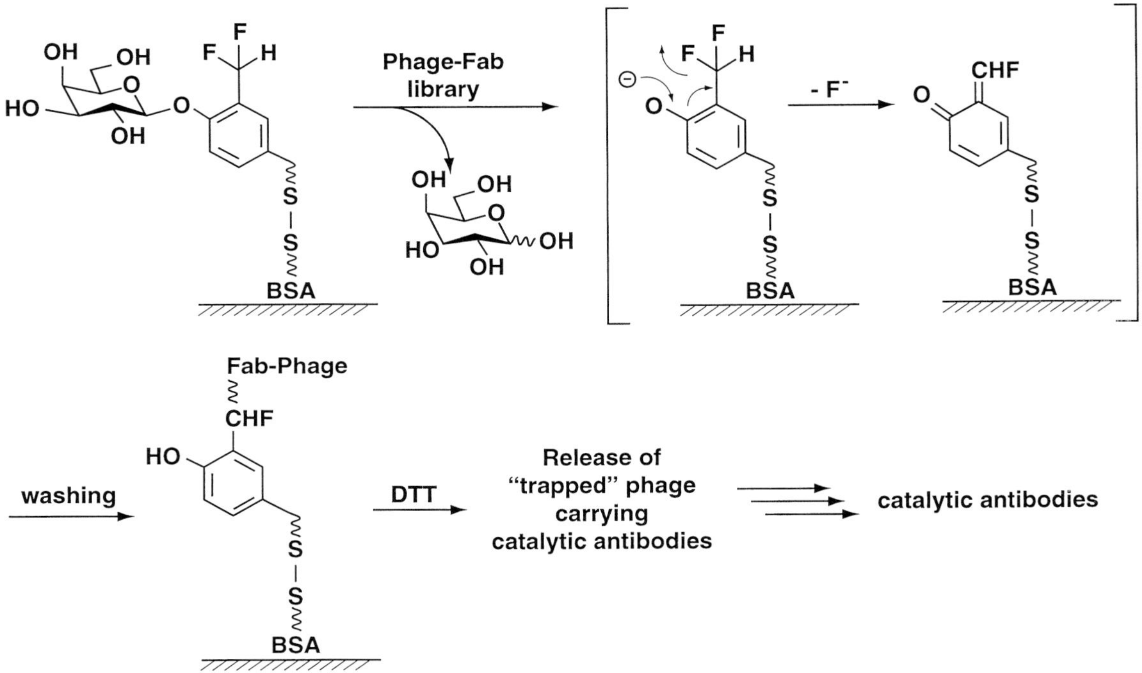

Fig. 30. The selection procedure using the reagent described in Figure 29. Reproduced from *Science*, **275,** 945–948 (1997).

driving force is improved function in terms of chemistry. Thus, they are selected for their ability to accomplish chemistry. The immune system also works on evolutionary principles, albeit in real time, but it selects for simple binding. Therefore, another way of looking at the evolution of the aldolase antibodies is that they have evolved from an alternative evolutionary system in which the selection parameters have been changed from simple binding to more complicated chemistry.

Finally, we can turn to how one should select antibody catalysts in the future. The central idea is to take a lesson from nature and link chemistry and replication. Up to this point, selection in chemical and biological experiments had a different meaning. In biology a selectable event must lead to improved function, whereas in chemistry it must allow discrimination between alternative chemical outcomes. However, the amalgamation of these two ideas has been made possible because of the construction of phage vectors in which antibodies and the genes encoding them are packaged, thereby linking replication and recognition in a clonal way. What had been missing, however, was the chemical component to link the ability to accomplish a chemical event with replication. This has now been done in a system in which a phage that accomplishes a chemical feat is selected (Janda et al., 1997). The purpose of the experiment was to make an efficient glycosidase. The necessary reagents for the experiment are shown in Figure 29. The hapten is an analogue that mimics the half chair conformation and oxocarbenium stereoelectronic features of the transition state of glycosidic bond hydrolysis, but the important construct is the panning reagent that selects for a phage that accomplishes chemistry. It has three key regions: (a) a glycosidic bond, (b) a leaving group that will become highly reactive if the glycosidic bond is broken, and (c) a linker that can be disrupted to recover a reactive phage. The basic scheme is shown in Figure 30 in which the panning reagent is linked to a solid support and the construct is exposed to a phage carrying potential antibody glycosidases on the surface. Any phage that hydrolyzes the glycosidic bond will liberate a highly reactive quinone methide that is unlikely to escape the binding pocket, thereby alkylating and immobilizing the phage on the solid support. All that is left to do then is to reduce the disulfide bond to liberate and grow the phage that accomplished the desired chemistry. Thus, replication and chemistry are linked! The glycosidases prepared in this way were found to be efficient (Fig. 31) and even capable of restoring function to glycosidase deficient *E. coli.* mutants (Fig. 32).

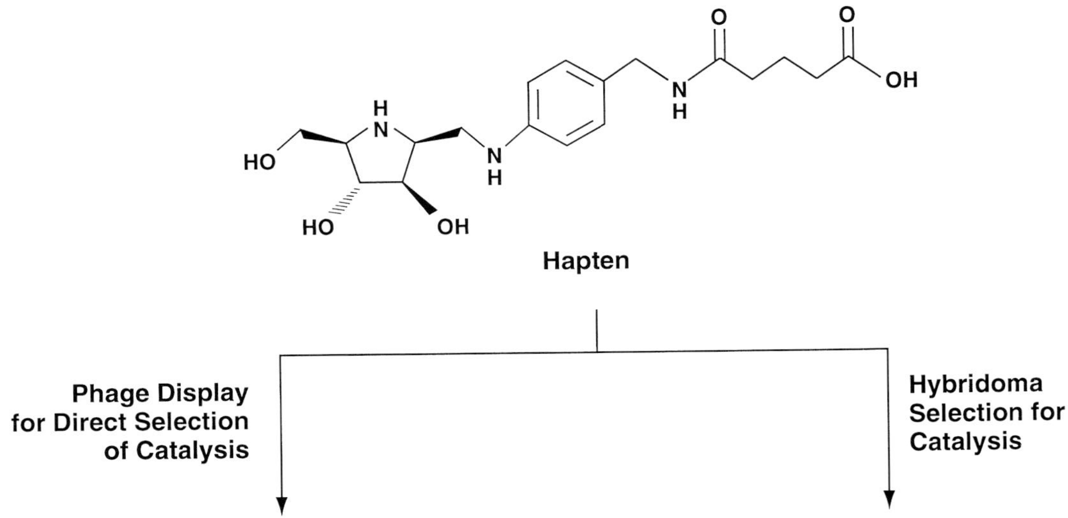

Phage Display for Direct Selection of Catalysis		Hybridoma Selection for Catalysis	
k_{cat}	$= 0.007\ min^{-1}$	k_{cat}	$= 10^{-5}\ min^{-1}$
K_M	$= 530\ \mu M$	K_M	$= 330\ \mu M$
k_{uncat}	$= 10^{-7}\ min^{-1}$	k_{uncat}	$= 10^{-7}\ min^{-1}$
k_{cat}/k_{uncat}	$= 7 \times 10^4$	k_{cat}/k_{uncat}	$= 100$

Fig. 31. Comparative rates of antibody glycosidases using usual hybridoma methods versus direct selection for catalysis in phage.

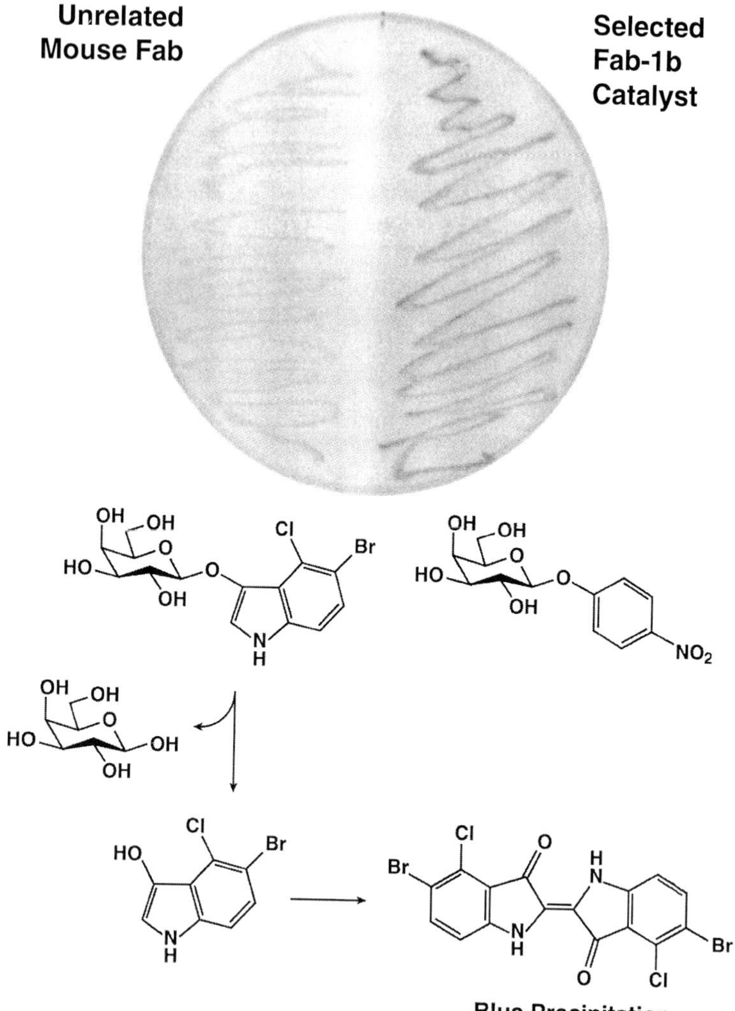
Unrelated
Mouse Fab
Selected
Fab-1b
Catalyst
Blue Precipitation

Figure 32.

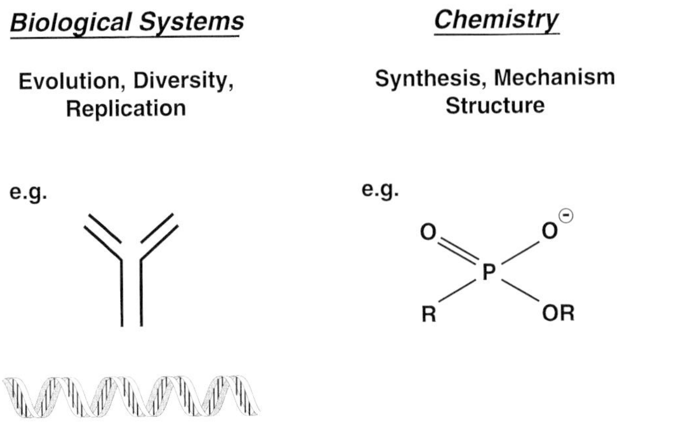

Fig. 33. The power of chemistry versus the power of biology. Each has unique strengths that when combined are extremely powerful to make new catalysts.

Chemical systems and biological systems each have unique powers. Biology has evolution, diversity, and replication. Chemistry has the power to understand reaction mechanisms, synthesize molecules of any kind, and know their structure. (Fig. 33). We have explored the potential of combining the unique powers of each to create new functions, in this case enzymes.

Fig. 32. Detection of catalytic antibody activity with a substrate-precipitating chromogenic assay: The DNA of Fab 1B was subcloned into the pWPY501 vector where protein expression is under the control of the T7 promoter. A positive selection for catalysis was confirmed by transformation of the pWPY501-1B plasmid into NoveBlue (DE3) cells, which are β-galacosidase-deficient. The cells were plated on Luria-Bertani (LB) agar plates containing 0.1 mM IPTG and 2% X-gal. (A) An unrelated mouse Fab elicited to a boronic acid hapten was cloned into pWPY501 and transformed into NoveBlue (DE3) cells (negative control). (B) The selected Fab-1B catalyst in NoveBlue (DE3) cells. Both cells containing the Fab genes (negative control and selected catalysts) were streaked onto a LB-IPTG-X gal plate from a single clone. The photograph was taken after incubation of the plate for 72 hours at 37°C. The compound corresponds to 5-bromo-4-chloro-3-indolyl-β-D-galactopyranoside (X-gal); the compound corresponds to p-nitrophenyl-β-D-galactopyranoside. Reproduced from *Science*, **275,** 945–948 (1997).

References

Amit, A.G., Mariuzza, R.A., Phillips, S.E., and Poljak, R.J. (1986). *Science* **233,** 747–753.

Arevalo, J.H., Taussig, M.J., and Wilson, I.A. (1993a). *Nature* **365,** 859–863.

Arevalo, J.H., Stura, E.A., Taussig, M.J., and Wilson, I.A. (1993a). *J. Mol. Biol.* **231,** 103–118.

Arevalo, J.H., Hassig, C.A., Stura, E.A., Sims, M.J., Taussig, M.J., and Wilson, I.A. (1994). *J. Mol. Biol.* **241,** 663–669.

Barbas, C.F., III, Heine, A., Zhong, G., Hoffmann, T., Gramatikova, S., Björnestedt, R., List, B., Anderson, J., Stura, E., Wilson, I.A., and Lerner, R.A. (1997). *Science* **278,** 2085–2092.

Gibson, A.L., Herron, J.N., He, X.-M., Patrick, V.A., Mason, M.L., Lin, J.-N., Kranz, D.M., Voss, E.W., Jr., and Edmundson, A.B. (1988). *Proteins Struc. Funct. Genet.* **3,** 155–160.

Haynes, M.R., Heine, A., and Wilson, I.A. (1996). *Isr. J. Chem.* **36,** 133–142.

Haynes, M.R., Stura, E.A., Hilvert, D., and Wilson, I.A. (1994). *Science* **263,** 646–652.

Hilvert, D., Carpenter, S.H., Nared, K.D., and Auditor, M.T. (1988). *Proc. Natl. Acad. Sci. U.S.A.* **85,** 4953–4955.

Israel Journal of Chemistry and references therein, **36** (2) (1996).

Jackson, D.Y., Jacobs, J.W., Sugasawara, R., Reich, S.H., Bartlett, P.A., and Schultz, P.G. (1988). *J. Am. Chem. Soc.* **110,** 4841–4842.

Janda, K.D., Lo, L.-C., Lo, C.-H.L., Sim, M.-M., Wang, R., Wong, C.-H., and Lerner, R.A. (1997). *Science* **275,** 945–948.

Rini, J.M., Schulze-Gahmen, U., and Wilson, I.A. (1992). *Science* **255,** 959–965.

Sheriff, S., Silverton, E.W., Padlan, E.A., Cohen, G.H., Smith-Gill, S.J., Finzel, B.C., and Davies, D.R. (1987). *Proc. Natl. Acad. Sci. U.S.A.* **84,** 8075–8079.

Smith-Gill, S.J., Mainhart, C., Lavoie, T.B., Feldmann, R.J., Drohan, W., and Brooks, B.R. (1987). *J. Mol Biol.* **194,** 713–724.

Stanfield, R.L., Fieser, T.M., Lerner, R.A., and Wilson, I.A. (1990). *Science* **248,** 712–719.

Wirsching, P., Ashley, J.A., Lo, C.-H.L., Janda, K.D., and Lerner, R.A. (1995). *Science* **270,** 1775–1782.

Wagner, J., Lerner, R.A., and Barbas, C.F., III (1995). *Science* **270,** 1797–1800.

Zhong, G., Hoffmann, T., Lerner, R.A., Danishefsky, S., Barbas, C.F., III (1997). *J. Am. Chem. Soc.* **119,** 8131–8132.

TUMOR NECROSIS FACTOR SIGNALING MECHANISMS

DAVID V. GOEDDEL

Tularik, Inc., South San Francisco, California

I. Introduction

The name *tumor necrosis factor* (TNF) was coined in 1975 to describe an activity produced by activated macrophages that was able to induce hemorrhagic necrosis of transplanted tumors in mice and kill certain transformed cell lines in culture (Carswell et al., 1975). Based on these and other early studies showing anti-cancer potential, several academic laboratories and biotechnology companies initiated efforts to molecularly characterize TNF. In 1984, human TNF was purified to homogeneity, and cDNA clones for human TNF were identified (Pennica et al., 1984). This work showed that TNF is initially synthesized as a precursor protein of 233 amino acids, from which the N-terminal 76 amino acids are removed. The mature form of TNF is a 52 kDa trimer (Eck and Sprang, 1989; Jones et al., 1989) consisting of identical 157 amino acid subunits.

The subsequent availability of pure, recombinant TNF permitted many studies of its biological properties. TNF is now known to be a highly pleiotropic cytokine with both beneficial and harmful actions. In addition to its ability to induce cell death, it is a major mediator of immune regulation and the inflammatory response. It also has antiviral, mitogenic, and many other activities (reviewed by Goeddel et al., 1986; Fiers, 1991). TNF entered clinical trials in the late 1980s as an anti-cancer agent, but the results to date have been disappointing. Clinical efforts have recently been directed at blocking the activity of TNF as it has become clear that inappropriate production of TNF is associated with a number of disease states such as septic shock and rheumatoid arthritis (reviewed by Fiers, 1991; Vassalli, 1992).

The Harvey Lectures, Series 92, pages 41–54

II. Two TNF Receptors

All of TNF's activities appear to be signaled through two distinct cell surface receptors that were identified by molecular cloning experiments in 1990 (Loetscher et al., 1990; Schall et al., 1990; Smith et al., 1990; Lewis et al., 1991). TNF receptor 1 (TNF-R1) and TNF receptor 2 (TNF-R2) are approximately 55 kDa and 75 kDa, respectively. The two receptors share 28% identity in their extracellular ligand-binding domains (ECDs). TNF binding to the ECD induces receptor trimerization (Banner et al., 1993), which initiates the signal transduction process (reviewed by Tartaglia and Goeddel, 1992; Vandenabeele et al., 1995). TNF-R1 and TNF-R2 belong to the large TNF receptor superfamily, currently numbering about 15 members, which is defined by the cysteine-rich architecture of the ligand binding domain (Smith et al., 1994). The ligands for most of these receptors share sequence similarity with TNF and are thus likely to be trimeric in structure (Smith et al., 1994).

The fact that TNF-R1 and TNF-R2 are co-expressed on most cell types complicated the assignment of various TNF activities to a particular receptor. However, transfection experiments, the use of receptor-specific antibodies, and experiments with gene knock-out mice have shown that TNF-R1 is the dominant signaling receptor (for reviews see Tartaglia and Goeddel, 1992; Vandenabeele et al., 1995) (Fig. 1). Among many other activities, TNF-R1 can induce apoptosis and activate the important transcription factor NF-κB. TNF-R2 can directly signal some activities such as NF-κB activation, and it also plays an auxiliary role by helping deliver low concentrations of TNF to TNF-R1 (Tartaglia et al., 1993a).

In contrast to the relatively well understood signaling events initiated by the receptor tyrosine kinases and hematopoietic cytokine receptors, little was known about post-receptor signaling mechanisms utilized by the TNF-R superfamily. The intracellular signaling domains (ICDs) of this receptor family do not specify enzymatic activity and are largely unrelated to each other. In fact, the primary amino acid sequences of the ICDs of the two TNF receptors are completely unrelated. However, the recent identification of several TNF-R-associated proteins and the determination of how they assemble into signaling complexes has advanced our knowledge considerably in this area. Common signaling themes for the entire superfamily have emerged from these studies of the two distinct TNF receptor complexes.

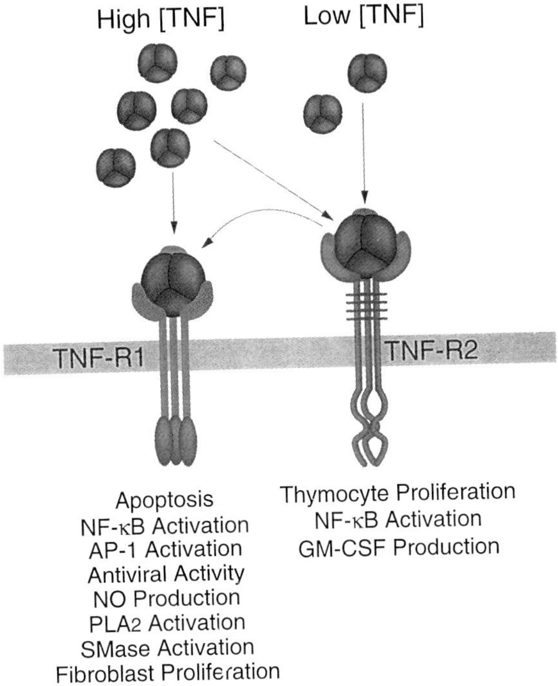

Fig. 1. Roles of the two TNF receptors in signal transduction. TNF-R1 and TNF-R2 share 28% sequence identity in their ECDs and are unrelated in their ICDs. TNF-R2 can "pass" low levels of TNF to TNF-R1 (Tartaglia et al., 1993a). Activities known to be directly signaled by TNF-R1 and TNF-R2 are listed under each receptor.

III. TNF Receptor 2 Signaling Complex

The mouse cytotoxic T cell line lacks TNF-R1. In response to murine TNF, the endogenous TNF-R2 signals proliferation and NF-κB activation in these cells. These same responses are seen in response to human TNF in stable transfectants of CT6 cells that express human TNF-R2. Analysis of various deletion mutants of human TNF-R2 in CT6 cells led to the identification of a 78 amino acid region in the middle of the ICD that were required for signaling (Rothe et al., 1994). Four distinct proteins of 45, 56, 66, and 68 kDa were found to bind to this region of TNF-R2 and were purified to homogeneity by affinity chromatography using a GST-TNF-R2 fusion protein (Rothe et al., 1994; 1995a).

The 45 and 56 kDa proteins were termed TRAF1 and TRAF2 for *T*NF *R*eceptor *A*ssociated *F*actors 1 and 2 (Rothe et al., 1994). Protein sequencing and cDNA cloning experiments indicated that these were novel proteins that contained a conserved C-terminal "TRAF" domain of about 230 amino acids. The TRAF domain can be divided into two subdomains, TRAF-N and TRAF-C, which share approximately 33% and 63% sequence identity, respectively, between TRAF1 and TRAF2. Additionally, the N-terminal half of TRAF2 contains a RING finger structure followed by five zinc fingers (Fig. 2).

Analysis of TRAF1 and TRAF2 revealed that these proteins can form both homo- and heterodimers through their TRAF domains, but that only TRAF2 interacts directly with TNF-R2 (Rothe et al., 1994). The interaction of TNF-R2 with TRAF2 occurs via the TRAF-C subdomain (Rothe et al., 1995b; Takeuchi et al., 1996). Thus, our initial purification of TRAF1 was not due to its direct interaction with TNF-R2, but was a consequence of the TRAF2-TRAF1 heterodimer associating with TNF-R2. Northern blot analysis suggests that TRAF2 expression is ubiquitous, whereas TRAF1 is expressed in a tissue-specific fashion. Therefore, depending on cell type, either a TRAF2–TRAF2 or a TRAF1–TRAF2 complex can be found associated with TNF-R2.

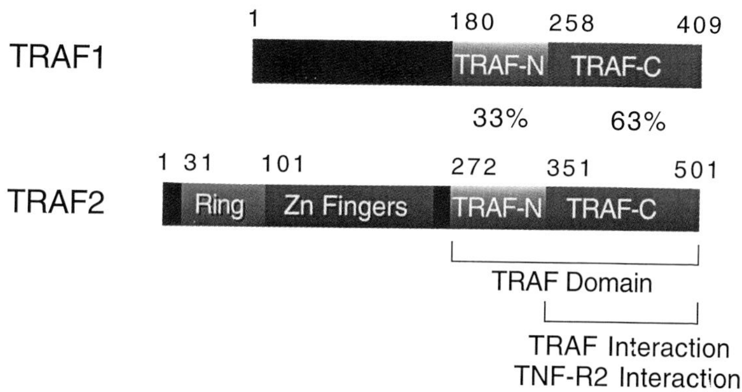

Fig. 2. Diagram of TRAF1 and TRAF2. The amino acid position of boundaries between the various domains of TRAF1 and TRAF2 are indicated above each bar. The degree of sequence identity between the TRAF-N and TRAF-C domains are also indicated.

The 68 and 66 kDa proteins that were affinity-purified using the GST-TNF-R2 fusion protein were given the names *c*ellular *I*nhibitor of *A*poptosis *P*rotein 1 and 2, or c-IAP1 and c-IAP2, respectively, because of their structural similarity to the baculoviral IAPs (Fig. 3) (Rothe et al., 1995a). Viral IAPs are only about 30 kDa but, like the mammalian c-IAPs, contain a C-terminal RING finger and N-terminal baculovirus IAP repeat (BIR) motifs. Since baculovirus IAPs prevent apoptosis of virally infected insect cells (Clem and Miller, 1994), it is likely that the mammalian c-IAPs are also involved in providing protection against cell death. However, the actual role(s) played by the c-IAPs in TNF signal transduction is not yet resolved. c-IAP1 mRNA is expressed in all tissues examined, whereas the expression of c-IAP2 is most prevalent in cells of the immune system (Rothe et al., 1995a).

Neither c-IAP1 nor c-IAP2 directly interact with TNF-R2. Instead, the BIR motifs of the cIAPs interact with the TRAF-N domains of TRAF1 and TRAF2. A TRAF1–TRAF2 heterodimer is required to recruit either c-IAP1 or c-IAP2 to the TNF-R2 complex (Fig. 4) (Rothe et al., 1995a). Thus, c-IAPs are absent from the TNF-R2 complex in cells lacking TRAF1.

Since TNF-R2 is capable of signaling NF-κB activation, we examined whether any of the four proteins in the TNF-R2 complex were responsible for transducing this signal. Each protein was separately expressed in human

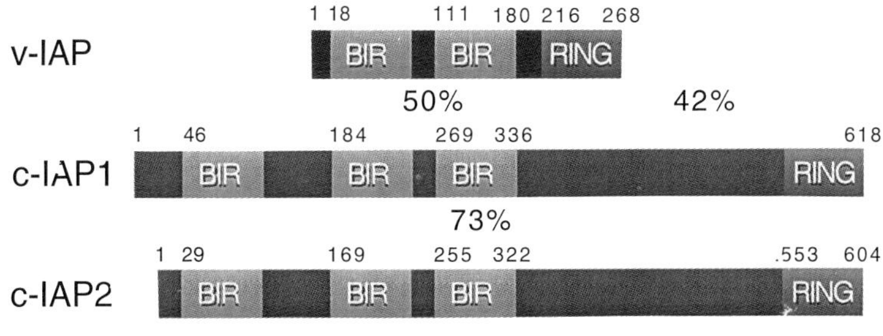

Fig. 3. Structural comparison of inhibitor of apoptosis proteins (IAPs). The positions of the conserved BIR motifs and RING finger structure of human c-IAP1 and c-IAP2 are shown in comparison to baculoviral IAP (v-IAP) from *Orgyia pseudotsugata* nuclear polyhedrosis virus. The two human c-IAPs are 73% identical in primary amino acid sequence. The BIR motifs and RING finger are 50% and 42% identical, respectively, between v-IAP and the c-IAPs.

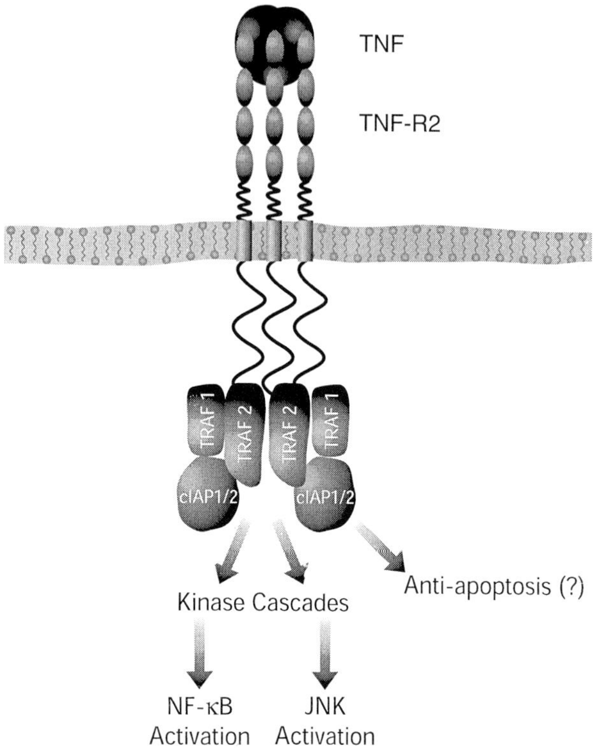

Fig. 4. The TNF-R2 signaling complex. TNF-induced trimerization of TNF-R2 results in the recruitment of TRAF1, TRAF2, c-IAP1, and c-IAP2 to the ICD of TNF-R2. Only TRAF2 directly contacts TNF-R2. TRAF2 is also responsible for the activation of kinase cascades leading to the activation of NF-κB and JNK. It is not yet known if the c-IAPs in this complex have anti-apoptotic activity.

293 cells in an attempt to mimic TNF-induced receptor aggregation. TRAF2 was found to potently activate NF-κB when overexpressed. Furthermore, a TRAF2 mutant lacking its N-terminal RING finger behaves as a dominant negative mutant and is a potent inhibitor of TNF-R2-mediated NF-κB activation (Rothe et al., 1995b) These results suggest that TRAF2 is required for activation of NF-κB by TNF-R2. Similarly, TRAF2 can activate the signaling pathway leading to activation of the Jun N-terminal kinase (JNK).

The TRAF2-mediated NF-κB and JNK activation pathways are not identical, as a dominant negative version of the kinase MEKK1 blocks JNK activation without affecting NF-κB activation (Liu et al., 1996).

TNF-R2, TRAF1, TRAF2, and the c-IAPs assemble into the TNF-R2 complex when overexpressed in a process that is TNF-independent (Rothe et al., 1995a). In order to determine whether the TRAFs and c-IAPs are associated with TNF-R2 under physiological conditions, endogenous TNF-R2 was immunoprecipitated from non-transfected CT6 cells before and after TNF treatment. The TNF-R2 immunocomplex was found to contain TRAF2 only following TNF treatment (M. Rothe, unpublished observations). Since TRAF2 is the only component of the TNF-R2 complex that directly contacts the receptor, it is clear that the signaling complex is assembled in a ligand-dependent process.

IV. TNF Receptor 1 Signaling Complex

The ICD of TNF-R1 contains a death domain of approximately 80 amino acids that is required for the signaling of two of TNF-R1's most important activities, apoptosis and NF-κB activation (Tartaglia et al., 1993b; Hsu et al., 1995). Yeast two-hybrid screening to search for proteins that interact with the ICD of TNF-R1 resulted in the identification TRADD. TRADD is a 34 kDa cytoplasmic protein that contains a C-terminal death domain sharing approximately 30% sequence identity with the death domain of TNF-R1 (Hsu et al., 1995). The interaction between TNF-R1 and TRADD occurs via their respective death domains, which are also capable of self-association. Transient overexpression of TRADD induces apoptosis and activates NF-κB, suggesting that TRADD is a TNF-R1 signal transducer. The signaling cascades for these two activities are at least partially distinct, as crmA, a pox virus cysteine protease inhibitor, blocks TRADD-induced apoptosis, yet has no effect on NF-κB activation (Hsu et al., 1995). TRADD and TNF-R1 are not pre-associated, but rapidly interact following TNF-mediated aggregation of the receptor (Hsu et al., 1996a).

To identify additional components of the TNF-R1 apoptosis and NF-κB activation pathways, two-hybrid screens for TRADD-interacting proteins were performed. These experiments yielded the proteins TRAF2, FADD, and RIP, each of which appears to play a role in TNF-R1 signaling (Hsu et al., 1996a; 1996b). In addition to these proteins, c-IAP1 is a component of the TNF-R1 complex, as it is present in immunoprecipitates of TNF-R1

(Shu et al., 1996). The proposed function(s) of each of these proteins in TNF-R1 signal transduction is discussed below.

A. TRAF2

TRADD interacts with the TRAF-C domain of TRAF2 and the death domain of TNF-R1 through its N- and C-termini, respectively (Hsu et al., 1996a). These observations suggested that TRADD might function as an adapter protein by simultaneously binding both TRAF2 and TNF-R1. This prediction was confirmed in both overexpression (Hsu et al., 1996a) and endogenous systems (Shu et al., 1996). Since TRAF2 is required for NF-κB activation by TNF-R2, its presence in the TNF-R1 complex suggested it may play a similar role in TNF-R1 signaling. Overexpression of a RING finger-deficient TRAF2 mutant effectively inhibited NF-κB activation by TNF-R1, but had no effect on apoptosis (Hsu et al., 1996a). These results show that TRAF2 plays a role in NF-κB activation by both TNF receptors, but is probably not involved in the signaling of apoptosis.

B. FADD

FADD was originally identified as a cytoplasmic death domain-containing protein of 26 kDa that interacts with the death domain of Fas, a member of the TNF receptor superfamily (Boldin et al., 1995; Chinnaiyan et al., 1995). FADD has an N-terminal death effector domain (DED) and a C-terminal death domain. The DED of FADD interacts with and recruits the cysteine protease caspase-8 to the Fas signaling complex where it can initiate Fas-mediated apoptosis (Boldin et al., 1996; Fernandes-Alnemri et al., 1996; Muzio et al., 1996).

We found that the death domains of TRADD and FADD interact efficiently. FADD does not interact directly with TNF-R1, but when all three proteins are overexpressed, TRADD is able to recruit FADD to the TNF-R1 complex (Hsu et al., 1996a). This finding suggests that death domains can form higher order complexes than dimers, and that FADD may be involved in TNF-R1-mediated signaling. Consistent with the latter hypothesis, we found that a deletion mutant of FADD lacking its N-terminal 79 amino acids is an potent inhibitor of TNF-R1-mediated cell death. This same DED-deleted FADD has no effect on TNF-R1-mediated NF-κB activation, indicating that FADD is not a part of this pathway (Hsu et al., 1996a). Despite considerable effort, we have been unable to demonstrate that FADD is a part of the TNF-R1 complex in non-transfected cells (H.-B. Shu and D.V. Goeddel, unpublished observations).

Thus, even though the dominant negative result provides strong suggestive evidence, it is possible that FADD is not required to deliver the TNF-R1 death signal. Alternatively, FADD's association with the TNF-R1 complex may be too transient to detect by the immunoprecipitation methods we have employed.

C. RIP

RIP is also a cytoplasmic protein that contains C-terminal death domain. It was originally isolated in a two-hybrid screen for Fas-interacting proteins (Stanger et al., 1995). Our characterization of RIP suggests that it does not interact with Fas in mammalian cells, and interacts poorly with TNF-R1 (Hsu et al., 1996b). However, TRADD interacts strongly with RIP and recruits RIP to the TNF-R1 complex very efficiently. Following TNF treatment, RIP is also found in the TNF-R1 complex in non-transfected cells (Hsu et al., 1996b). The N-terminal 300 amino acids of the 76 kDa RIP comprise a kinase domain, and RIP has serine/threonine protein kinase activity (Hsu et al., 1996b). Both the kinase domain and the intermediate domain of RIP interact with TRAF1, TRAF2, and TRAF3.

Like TRADD, RIP can activate NF-κB and induce apoptosis when overexpressed (Hsu et al., 1996b). RIP overexpression also activates JNK (Liu et al., 1996). Mutational analysis shows that both NF-κB and JNK activating functions reside in the kinase and intermediate domains, whereas the cell death activity is confined to the death domain (Hsu et al., 1996b; Liu et al., 1996). When expressed by itself, the C-terminal death domain of RIP is a very effective dominant negative inhibitor of TNF-induced NF-κB and JNK activation. Taken together, these data support a model in which RIP is required for TNF-mediated activation of NF-κB and JNK, and may play a role in TNF-mediated apoptosis.

D. c-IAP1

Following TNF treatment, immunoprecipitates of endogenous TNF-R1 from HeLa and U937 cells were found to contain c-IAP1 (Shu et al., 1996). Prior to TNF-treatment, c-IAP1 is not associated with the receptor complex, but is found complexed with TRAF2, suggesting that perhaps TRAF2 recruits c-IAP1 to TNF-R1. As predicted, reconstitution experiments show that c-IAP1 recruitment to TNF-R1 requires both TRADD and TRAF2 (Shu et al., 1996). The role of c-IAP1 in TNF-R1 signaling remains uncertain, but overexpression of c-IAP1 and c-IAP2 together seems to block TRADD-mediated apoptosis (H.-B. Shu and D.V. Goeddel, unpublished observations).

E. Assembly of the TNF-R1 Signaling Complex

In our model for the TNF-R1 signaling complex, TNF-R1 does not associate with other signaling proteins prior to its interaction with TNF. Following TNF treatment, TNF-R1 trimerizes and its self-associating death domains stabilize the aggregated receptor. TRADD, which exists as either a monomer or dimer, can now associate with the death domain of TNF-R1. This interaction allows TRADD to become an adapter protein that is now able to bind and recruit other signaling proteins. The N-terminus of TRADD re-

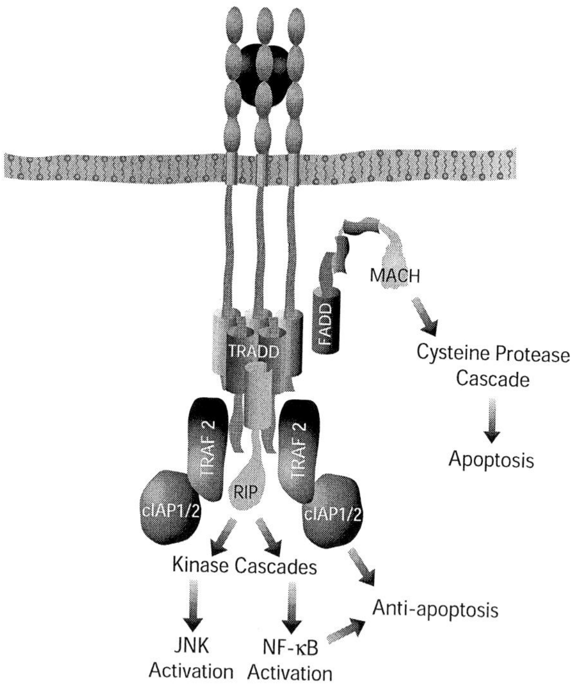

Fig. 5. The TNF-R1 signaling complex. TNF-induced trimerization of TNF-R1 results in the recruitment of TRADD, TRAF2, RIP, c-IAP1, c-IAP2 and perhaps FADD to the ICD of TNF-R1. RIP and TRAF2 can activate kinase cascades leading to the activation of NF-κB and JNK. NF-κB and the c-IAPs generate anti-apoptotic signals. TRADD, RIP, and probably FADD and MACH (caspase 8) activate the apoptotic pathway.

cruits TRAF2/c-IAP1 and the death domain of TRADD recruits RIP, and perhaps FADD, to the TNF-R1 complex. A model of the assembled complex indicating how it can transduce some of the signals is shown in Figure 5.

V. TRAF PROTEINS

Four additional TRAF proteins have been described since our initial purification of TRAF1 and TRAF2. TRAF3 was isolated by two-hybrid cloning on the basis of its interaction with CD40 (Hu et al., 1994; Cheng et al., 1995; Sato et al., 1995) and the Epstein-Barr virus protein LMP-1 (Mosialos et al., 1995). TRAF4 is a nuclear protein that was originally identified on the basis of amplification and overexpression of its gene in breast carcinomas (Regnier et al., 1995). TRAF5 was identified by degenerate PCR (Nakano et al., 1996) and by its interaction with CD40 (Ishida et al., 1996). TRAF6 was found in a search of public DNA sequence databases (Cao et al., 1996). Each of these additional TRAFs is similar to TRAF2 in that they contain N-terminal RING and zinc finger structures (Fig. 6).

TRAFs 1, 2, 3, and 5 interact with, and appear to be general signal transducers for, multiple members of the TNF receptor superfamily (above references, Lee et al., 1996; Gedrich et al., 1996). In each case, the interaction

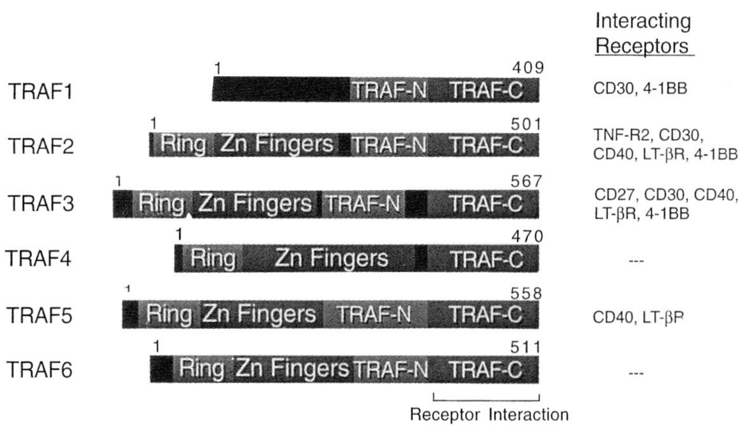

Fig. 6. Structural comparison of TRAF proteins. The members of the TNF receptor superfamily that are known to directly interact with each TRAF are shown.

requires the TRAF-C domain of the TRAF protein and a short continuous epitope of the receptor ICD. No proteins that interact with TRAF4 have yet been reported. TRAF6, which is only about 30% identical to the other TRAF proteins in the TRAF-C domain (Cao et al., 1996), may not interact with members of the TNF receptor superfamily.

Transfection experiments have shown that TRAF2 (Rothe et al., 1995b), TRAF5 (Ishida et al., 1996; Nakano et al., 1996), and TRAF6 (Cao et al., 1996) can activate NF-κB when overexpressed. These same three TRAFs can also activate JNK (Liu et al., 1996; H. Hsu, M. Rothe, and D.V. Goeddel, unpublished observations). Experiments with dominant negative TRAFs suggest that TRAF2 may be a general activator of NF-κB for the TNF receptor superfamily (Rothe et al., 1995b; Lee et al., 1996), whereas TRAF5 involvement may be more restricted (Nakano et al., 1996).

Recent experiments support a role for TRAF6 in signal transduction by a distinct class of receptors. A truncated version of TRAF6 lacking the entire zinc binding region failed to inhibit TNF signaling, but acted as an inhibitor of NF-κB activation triggered by the cytokine interleukin 1 (IL-1) (Cao et al., 1996). This result suggested that TRAF6 may associate with members of the IL-1 receptor superfamily. However, we have been unable to demonstrate such an interaction. Instead, TRAF6 appears to function by interacting with the IL-1 receptor-associated kinase (IRAK) after IRAK has been activated and released from the IL-1 receptor complex (Cao et al., 1996). How TRAFs function to activate kinase signaling cascades that result in the activation of NF-κB and JNK is currently under investigation.

VI. Conclusions

An understanding of the mechanisms by which the cytokine TNF signals so many diverse responses has begun to emerge from studies of TNF's two receptors and their associated signaling proteins. Various combinations of three distinct families of proteins, the TRAFs, c-IAPs, and death domain-containing proteins, are recruited to the receptors in a ligand-dependent process. The actual composition of a signaling complex in a given cell type is likely to influence important cellular decisions, such as whether the cell will undergo apoptosis or activate genes involved in the inflammatory response. The lessons learned from study of the TNF receptor complexes should provide a paradigm for signal transduction mechanisms utilized by other members of the expanding TNF receptor superfamily.

ACKNOWLEDGMENTS

I would like to thank the various members of my laboratory who, over the last several years, have contributed to the ideas, experiments, and interpretations described in this lecture. I am particularly indebted to Hailing Hsu and Mike Rothe, whose efforts were key to the elucidation of the signaling proteins utilized by TNF-R1 and TNF-R2, respectively.

REFERENCES

Banner, D., D'Arcy, A., Janes, W., Gentz, R., Schoenfeld, H.-J., Broger, C., Loetscher, H., and Lesslauer, W. (1993). *Cell* **73,** 431–455.

Boldin, M.P., Mett, I.L., Varfolomeev, E.E., Chumakov, I., Shemer-Avni, Y., Camonis, J.H., and Wallach, D. (1995). *J. Biol. Chem.* **270,** 387–391.

Boldin, M.P., Goncharov, T.M., Goltsev, Y.V., and Wallach, D. (1996). *Cell* **85,** 803–815.

Cao, Z., Xiong, J., Takeuchi, M., Kurama, T., and Goeddel, D.V. (1996). *Nature* **383,** 443–446.

Carswell, E.A., Old, L.J., Kassel, R.L., Green, S., Fiore, N., and Williamson, B. (1975). *Proc. Natl. Acad. Sci. U.S.A.* **72,** 3666–3670.

Cheng, G., Cleary, A.M., Ye, Z., Hong, D.I., Lederman, S., and Baltimore, D. (1995). *Science* **267,** 1494–1498.

Chinnaiyan, A.M., O'Rourke, K., Tewari, M., and Dixit, V.M. (1995). *Cell* **81,** 505–512.

Clem, R.J. and Miller, L.K. (1994). *In* "Apoptosis II: The Molecular Basis of Apoptosis in Disease," pp. 89-110, Cold Spring Harbor Laboratory Press, New York.

Eck, M.J., and Sprang, S.R. (1989). *J. Biol. Chem.* **264,** 17595–17605.

Fernandes-Alnemri, T., Armstrong, R.C., Krebs, J., Srinivasula, S.M., Wang, L., Bullrich, F., Fritz, L.C., Trapani, J.A., Tomaselli, K.J., Litwack, G., and Alnemri, E.S. (1996). *Proc. Natl. Acad. Sci. U.S.A.* **93,** 7464–7469.

Fiers, W. (1991). *FEBS Lett.* **285,** 199–212.

Gedrich, R.W., Gilfillan, M.C., Duckett, C.S., Van Dongen, J.L., and Thompson, C.B. (1996). *J. Biol. Chem.* **271,** 12852–12858.

Goeddel, D.V., Aggarwal., B.B., Gray, P.W., Leung, D.W., Nedwin, G.E., Palladino, M.A., Patton, J.S., Pennica, D., Shepard, H.M., Sugarman, B.J., and Wong, G.H.W. (1986). *In* "Cold Spring Harbor Symposia on Quantitative Biology," pp. 597-609, Cold Spring Harbor Laboratory Press, New York.

Hsu, H., Xiong, J., and Goeddel, D.V. (1995). *Cell* **81,** 495–504.

Hsu, H., Shu, H.B., Pan, M.G., and Goeddel, D.V. (1996a). *Cell* **84,** 299–308.

Hsu, H., Huang, J., Shu, H.B., Baichwal, V., and Goeddel, D.V. (1996b). *Immunity* **4,** 387–396.

Hu, M.H., O'Rourke, K., Boguski, M.S., and Dixit, V.M. (1994). *J. Biol. Chem.* **269,** 30069–30072.

Ishida, T., Tojo, T., Aoki, T., Kobayashi, N., Ohishi, T., Watanabe, T., Yamamoto, T., and Inoue, J.-I. (1996). *Proc. Natl. Acad. Sci. U.S.A.* **93,** 9437–9442.

Jones, E.Y., Stuart, D.I. and Walker, N.P.C. (1989). *Nature* **338,** 225–228.

Lee, S.Y., Lee, S.Y., Kandala, G., Liou, M.-L., Liou, H.-C., and Choi, Y. (1996). *Proc. Natl. Acad. Sci. U.S.A.* **93,** 9699–9703.

Lewis, M., Tartaglia, L.A., Lee, A., Bennett, G.L., Rice, G.C., Wong, G.H.W., Chen, E.Y., and Goeddel, D.V. (1991). *Proc. Natl. Acad. Sci. U.S.A.* **88,** 2830–2834.

Liu, Z., Hsu, H., Goeddel, D.V., and Karin, M. (1996). *Cell* **87,** 565–576.

Loetscher, H., Pan, Y.-C. E., Lahm, H.W., Gentz, R., Brockhaus, M., Tabuch, H., and Lesslauer, W. (1990). *Cell* **61,** 351–359.

Mosialos, G. Birkenbach, M., Yalamanchili, R., Van Arsdale, T., Ware, C., and Kieff, E. (1995). *Cell* **80,** 389–399.

Muzio, M. Chinnaiyan, A.M., Kischkel, F.C., O'Rourke, K., Shevchenko, A., Ni, J. (1996). *Cell* **85,** 817–827.

Nakano, H., Oshima, H., Chung, W., Williams-Abbott, L., Ware, C.F., Yagita, H., and Okumura, K. (1996). *J. Biol. Chem.* **271,** 14661–14664.

Pennica, D., Nedwin, G.E., Hayflick, J.S,. Seeburg, P.H., Derynck, R., Palladino, M.A., Kohr, W.J., Aggarwal, B.B., and Goeddel, D.V. (1984). *Nature* **312,** 724–729.

Regnier, C.H., Tomasetto, C., Moog-Lutz, C., Chenard, M.-P., Wendling, C., Basset, P., and Rio, M.-C. (1995). *J. Biol. Chem.* **270,** 25715–25721.

Rothe, M., Wong, S.C., Henzel, W.J., and Goeddel, D.V. (1994). *Cell* **78,** 681–692.

Rothe, M., Pan, M.-G., Henzel, W.J., Ayres, T.A., and Goeddel, D.V. (1995a). *Cell* **83,** 1243–1252.

Rothe, M., Sarma, V., Dixit, V.M., and Goeddel, D.V. (1995b). *Science* **269,** 1424–1427.

Sato, T., Irie, S., and Reed, J.C. (1995). *FEBS Lett.* **358,** 113–118.

Schall, T.J., Lewis, M., Koller, K.H., Lee, A., Rice, G.C., Wong, G.H.W., Gatanaga, T., Granger, G.A., Lentz, R., Raab, H., Kohr, W.J., and Goeddel, D.V. (1990). *Cell* **61,** 361–370.

Shu, H.-B., Takeuchi, M., and Goeddel, D.V. (1996). *Proc. Natl. Acad. Sci. U.S.A.* **93,** 13973–13978.

Smith, C.A., Davis, T., Anderson, D., Solam, L., Beckmann, M.P., Jerzy, R., Dower, S.K., Cosman, D., and Goodwin, R.G. (1990). *Science* **248,** 1019–1023.

Smith, C.A., Farrah, T., and Goodwin, R.G. (1994). *Cell* **76,** 959–962.

Stanger, B.Z., Leder, P., Lee, T.H., Kim, E., and Seed, B. (1995). *Cell* **81,** 513–523.

Takeuchi, M., Rothe, M., and Goeddel, D.V. (1996). *J. Biol. Chem.* **271,** 19935–19942.

Tartaglia, L.A., and Goeddel, D.V. (1992). *Immunol. Today* **13,** 151–153.

Tartaglia, L.A., Pennica, D., and Goeddel, D.V. (1993a). *J. Biol. Chem.* **268,** 18542–18548.

Tartaglia, L.A., Ayres, T.M., Wong, G.H.W., and Goeddel, D.V. (1993b). *Cell* **74,** 845–853.

Vandenabeele, P., Declercq, W., Beyaert, R., and Fiers, W. (1995). *Trends Cell Biol.* **5,** 392–399.

Vassalli, P. (1992). *Ann. Rev. Immunol.* **10,** 411–452.

THE CENTRAL ROLE OF A CDK IN CONTROLLING THE FISSION YEAST CELL CYCLE

PAUL NURSE

Imperial Cancer Research Fund, Lincoln's Inn Fields, London, England

> Nature is nowhere accustomed more openly to display her secret mysteries than in cases where she shows traces of her workings apart from the beaten path; nor is there any better way to advance the proper practice of medicine than to give our minds to the discovery of the usual law of Nature by careful investigation of rarer forms of disease. For it has been found in almost all things, that what they contain of useful or applicable nature is hardly perceived unless we are deprived of them, or they become deranged in some way. *(From a letter written six weeks before William Harvey's death in 1657.)*

I. INTRODUCTION

The cell cycle is the process by which cells reproduce themselves, and controls acting over the cell cycle ensure an orderly progression through this reproductive process. An important element controlling cell cycle progression in eukaryotes is the cyclin dependent kinase (CDK) family of protein kinases, which regulate passage through the major events of the cell cycle in all eukaryotes. In this paper attention is focused on the simple unicellular eukaryote fission yeast *Schizosaccharomyces pombe*. Emphasis is placed on the role of cdc2p, the catalytic core of the CDK protein kinase that controls the cell cycle in fission yeast. Cell cycle controls are more elaborate in multicellular eukaryotes, but focusing on a single model system has the advantage of providing a more coherent and complete description of how the controls operate than is yet possible in more complex systems. Also, understanding generated with a simple model such as fission yeast can be of use in unravelling the more elaborate and redundant controls operative in more complicated Metazoan cells.

The Harvey Lectures, Series 92, pages 55–64

II. Cell Cycle Controls

The cell cycle is often defined as the period in the life of a growing cell between the formation of the cell by the division of its mother and the time when the cell itself divides to form two daughters. This is a useful definition because it places emphasis on the fact that the cell cycle usually occurs only when cells are already in an active state of growth, and thus discussion about cell cycle control is not confused with cell growth control. The latter regulates the transition from a non-growing to a growing state. In a free-living unicellular organism such as fission yeast, growth controls are concerned with the availability of essential nutrients in the growth medium, while in a mammalian cell growth, controls are concerned with growth factor signals that reflect the social interactions of cells operative in a multicellular organism. These growth controls are of great importance but will not be dealt with here. Consideration will be confined to the controls regulating progression through the cell cycle of a cell that is already actively growing.

For the cell cycle to be successful it is necessary that at cell division the two daughters each receive a full complement of all the components necessary for cell survival. Of particular importance is the genome, because each new daughter cell should contain a complete set of genes. Two events present in all normal eukaryotic cell cycles are essential to achieve this end. These are S-phase (S), when the DNA making up the genome is replicated, and M-phase (M) or mitosis in the mitotic cell cycle, when the replicated chromosomes are segregated into the two daughter cells. An important aspect of cell cycle control is concerned with regulating the onset of these two events such that they occur in the correct sequence of S-phase followed by mitosis, and only once in each cell cycle. Therefore much of the work considered here is concerned with elucidating the mechanisms that control onset of these two events.

Precisely replicating and accurately segregating the genome is a complex process that can go wrong. Cell cycle checkpoint controls ensure that a cell does not divide with a partly or incorrectly replicated genome and that mistakes do not occur during chromosomal segregation. If a mistake occurs during S-phase that blocks DNA replication, then the onset of mitosis is also blocked. Similarly, if chromosome segregation is defective, then cell division is blocked. These blocks give time for the mistakes and defects to be corrected, after which the blocks are relieved and normal cell cycle progression is allowed. Checkpoint controls are crucial for the fidelity of the genomic reproductive process and therefore also form an important part of cell cycle control. Also of relevance in this regard is the mechanism that ensures that there is only one S-phase per cell cycle. Failure in this mechanism would lead to an increase in cell ploidy, and although each daughter would still receive a full complement of the genes, the ploidy changes would be detrimental for both proper development and sexual reproduction. These controls, which are closely related to those regulating the onset of the events of S-phase and mitosis, maintain genomic stability, which is essential for the survival of both unicellular and multicellular organisms.

Elucidation of these controls is difficult because of their complexity, but is much assisted by a genetic approach. Specific mutants disrupt single elements acting in the controls, and careful study of the subsequent mutant phenotype or pathology is revealing about the normal processes involved. When this is combined with molecular genetics allowing the cloning of the relevant genes, it is then possible to work out the molecular mechanisms underlying the controls. Fission yeast is a genetically amenable organism that is particularly suited for such a genetic analysis of cell cycle controls (Nurse, 1975; Nurse, et al., 1976). Work with this organism has been illuminating about most of the controls outlined above. In particular it has drawn attention to the *cdc2* gene (encoding cdc2p) as an important element in the controls (Nurse, 1990), and the mode of action and regulation of cdc2p will now be summarized.

III. Action and Regulation of cdc2p

As mentioned above, cdc2p is a catalytic protein kinase core (Simanis and Nurse, 1986). It is 34 kD in size and must be complexed with a cyclin regulatory subunit to become active. There are four cyclins in fission yeast; three of these are B-cyclins, cdc13p (Hagan et al., 1988), cig1p, and cig2p, all of which have roles in the cell cycle. The fourth is puc1p, which is closely related to the CLN cyclins of budding yeast (Forsburg and Nurse, 1991). In fission yeast puc1p has no clearly identified role in the cell cycle, but does have a minor role in controlling conjugation and meiosis (Forsburg and Nurse, 1994). As well as this requirement for cyclin binding, cdc2p needs to be phosphorylated on threonine T167 for active protein kinase activity (Gould et al., 1991). This T167 phosphorylation may stabilize cyclin binding.

The cdc2p protein kinase activity is subject to three further types of regulation. The first is an inhibitory phosphorylation of tyrosine Y15 located in the active site of the enzyme (Gould and Nurse, 1989). When phosphorylated on Y15 the protein kinase activity is reduced although not eliminated. Specific protein kinases (primarily wee1p) (Russell and Nurse, 1987) and protein phosphatases (primarily cdc25p) (Russell and Nurse, 1986; Gould et al., 1990) regulate the phosphorylation state of Y15. The second type of regulation is association of the cdc2p CDK with a specific CDK inhibitor rum1p (Moreno and Nurse, 1994; Correa-Bordes and Nurse, 1995). This association completely inhibits the cdc2p-cdc13p CDK protein kinase activity at least *in vitro,* but is less effective against the cdc2p-cig1p and cdc2p-cig2p CDKs. The third type of regulation is controlled proteolysis of the cyclin component of the cdc2p CDK. Proteolysis of cdc13p is brought about by the proteosome, although rum1p may play a role in efficiently directing cdc13p to the proteosome. All three types of regulation are used at different stages of cell cycle regulation in fission yeast.

IV. Controlling Mitotic Onset

The best understood cell cycle transition in fission yeast is the control acting at mitotic onset. This is brought about by the cdc2p-cdc13p CDK, although cdc2-cig1p CDK may play a very minor role. The complex of cdc2p with cdc13p begins to form around S-phase, and sufficient complex for adequate mitotic protein kinase activity is present by the first part of G2 (Moreno et al., 1989). The main control determining the timing of mitotic onset involves Y15 phosphorylation (Gould and Nurse, 1989). As soon as the cdc2p-cdc13p complex is formed, Y15 becomes phosphorylated primarily by the wee1p protein kinase, although the mik1p protein kinase also plays a contributory role. This leads to a low level of cdc2p-cdc13p protein kinase activity from S-phase to G2 during the cell cycle. The Y15 phosphate is removed primarily by the cdc25p protein phosphatase, although the pyp3p protein phosphatase also plays a contributory role. When Y15 is dephosphorylated, a high level of cdc2p-cdc13p protein kinase is formed, which brings about mitosis. Thus activation of the cdc2p-cdc13p CDK at the end of G2 depends mainly on the balance between the activities of the wee1p protein kinase and the cdc25p protein phosphatase. Activation occurs when the cell attains a critical mass (Nurse, 1975), and removing wee1p (Russell and Nurse, 1987) or over-expressing cdc25p (Russell and Nurse, 1986) advances cells into mitosis at a reduced cell mass.

When the cdc2p-cdc13p CDK is fully activated, mitosis takes place. A number of key substrates required for the major events of mitosis are thought to become phosphorylated at this time (Moreno and Nurse, 1990). These include formation of a mitotic spindle, chromosome condensation, and changes in the nuclear envelope that occurs as the nucleus extends during mitosis. Little work has been done on these substrates, but cdc2p has been found in the spindle pole body (of obvious relevance for forming the mitotic spindle), and when a vertebrate lamin (a component of the nuclear envelope) is expressed in fission yeast, it is phosphorylated by cdc2p-cdc13p and becomes dispersed during mitosis (Enoch et al., 1991). To get out of mitosis, the cdc2p-cdc13p CDK activity must be much reduced. This is brought about by specific cdc13p proteolysis and requires the action of nuc2p and cut9p, two components of the proteosome. When these components are defective, cells block in mitosis because of their failure to degrade cdc13p, and as a consequence a high level of cdc2p-cdc13p protein kinase activity is maintained, which blocks mitotic exit. Only when cdc2p-cdc13p activity is at a low level can cells exit mitosis and complete the cell cycle.

V. Controlling S-Phase Onset

Onset of S-phase is less well understood. The primary cyclin involved is cig2p, and cdc2p-cig2p CDK protein kinase activity rises to a peak at the G1/S boundary. If cdc2p is inactive, then cells fail to enter S-phase, but the mechanisms controlling activation and inactivation of cdc2p-cig2p have yet to be elucidated.

cig2p is not the only B-cyclin that can act at the G1/S boundary. Both cig1p and cdc13p can substitute for cig2p if the latter cyclin is not present. Usually cdc2p-cig2p is activated earlier in the cell cycle, and so neither cig1p nor cdc13p normally have any function at G1/S. However, if the *cig2* gene is deleted, then either cig1p or cdc13p can bring about S-phase (Fisher and Nurse, 1996). When both *cig1* and *cig2* are deleted, the mitotic-cyclin cdc13p becomes essential for onset of both S-phase and mitosis. Activity associated with cdc2p-cdc13p increases during the cell cycle, first bringing about S-phase and then mitosis because the latter event requires a greater level of activity.

The molecular mechanism by which cdc2p brings about S-phase is likely to involve cdc18p. This protein plays a key role for the initiation of DNA replication in fission yeast (Kelly et al., 1993; Nishitani and Nurse, 1995). When *cdc18* is deleted, onset of S-phase is blocked, and when *cdc18* is over-expressed, multiple rounds of DNA replication occur even in the absence of continued protein synthesis. This identifies cdc18p as a major rate-limiting step for the initiation of DNA replication. cdc18p complexes with orp1p, an orc1p-related protein thought to be part of an origin recognition complex (ORC) associated with origins of replication, and with cdc21p, a mcm4p-related protein (Grallert and Nurse, 1996). At present, speculation on the modes of cdc18p action must be tentative, but it could be imagined that cdc18p is required for S-phase onset at a point downstream of cdc18p accumulation, and so cdc2p-cig2p CDK protein kinase activity may act at the point where cdc18p activates ORCs.

VI. Blocking M until S is Complete

An important cell cycle checkpoint control is the block over mitosis that occurs when S-phase is incomplete (Enoch and Nurse, 1991). This control is revealed when DNA replication is blocked with the inhibitor hydroxyurea, which blocks mitotic onset (Enoch and Nurse, 1990; Enoch et al., 1992). The control works through cdc2p because mutants that reduce cdc2p Y15 phosphorylation fail to block mitosis even though S-phase is incomplete. Various genes have been identified that are required for this checkpoint blocking signal. Mutants in these genes lead to *h*ydroxy*u*rea *s*ensitivity and so the genes implicated are called *hus* genes. Many *hus* genes are also radiation-sensitive and are found to be identical to previously identified *rad* genes. This has led to the view that there is at least some overlap between the mechanisms blocking mitosis due to failures to complete DNA replication and due to DNA damage. The present view is that blocks in DNA replication and damage to DNA send signals that are communicated to the cdc2p-cdc13p mitotic CDK by *hus/rad* gene pathway. Although both checkpoints share some common features, other gene functions may be unique to each pathway.

cdc18p, together with a specific subset of proteins required for the initiation of DNA replication, is crucial for sending the signal that S-phase is incomplete. When *cdc18* is deleted, cells fail to initiate S-phase and also fail to block the subsequent mitosis. This suggests that cdc18p is required both to initiate DNA replication and also for sending the signal that DNA replication is in process. Other replication proteins that behave in a similar way include orp1p and DNA polymerase alpha (D'Urso et al., 1995). All three proteins are required at an early stage in the initiation of DNA replication. Loss of other replication proteins acting later in the initiation process, including DNA polymerase delta and epsilon and PCNA, block S-phase and also block mitotic onset. This indicates that at these later mutant blocks the checkpoint signal monitoring that DNA replication is in process has been sent. This can be understood if the checkpoint signal is sent by the formation of replicative complexes acting at the initiation of DNA replication. These are formed at the onset of S-phase, and their presence signals that DNA replication is in process. All these replication complexes are consumed during the process of DNA replication, and so at the end of S-phase no further inhibitory signal is sent and cells can undergo mitosis. Deleting genes that block late in the replication process (e.g., DNA polymerase delta) does not affect formation of the replicative complexes, and so mitosis is blocked. In contrast, deleting genes that block early (e.g., cdc18p) prevents the formation of the complexes, and so no inhibitory signal is sent and cells enter mitosis.

An early G1 cell has also not completed S-phase and should not undergo mitosis, and yet it has not yet formed any replicative complexes and so cannot block onset of mitosis by the mechanism described above. During this phase of the fission yeast cell cycle it is rum1p that prevents mitotic onset (Moreno and Nurse, 1994; Correa-Bordes and Nurse, 1995). When a cell completes mitosis and enters G1, rum1p levels rise, inhibiting any cdc2p-cdc13p CDK present and promoting cdc13p protein turnover. This mechanism ensures that early G1 cells do not enter mitosis. However, if *rum1* is deleted, cells blocked in early G1 will proceed to mitosis even though S-phase has not taken place. *rum1*-deleted cells blocked in S-phase using hydroxyurea do not proceed to mitosis because the Y15 phosphorylation checkpoint control is still intact in these cells.

VII. Blocking S until M is Complete

When a fission yeast cell is arrested in G2, it does not re-initiate another S-phase. In other words, S-phase can only take place when the mitosis of the previous cell cycle is completed, that is, there can only be one S-phase per cell cycle. This checkpoint control also involves the cdc2p-cdc13p CDK that inhibits re-initiation of DNA replication during the G2 phase of the cell cycle. When *cdc13* is deleted (Hayles et al., 1994) or the specific cdc2p-cdc13p CDK inhibitor rum1p is over-expressed (Moreno and Nurse, 1994), there is a very low level of cdc2p-cdc13p CDK protein kinase activity present in the cell. As a consequence, G2-arrested cells re-initiate DNA replication and so cell ploidy increases.

The advantage of this inhibitory control mechanism is that it explains why there is only one S-phase during each cell cycle. In the normal cell cycle, cdc2p-cdc13p protein kinase activity has to be reduced to a very low level before a cell can exit mitosis. Such reduction would not only lead to mitotic exit but also to release from the block acting over the re-initiation of DNA replication. Thus the cdc2p-cdc13p mitotic CDK must have two roles during G2. It prepares the cell for mitosis, the legitimate event for a G2 cell, and it prevents the cell from undergoing S-phase, the illegitimate event for a G2 cell. Upon exit from mitosis and entering G1, the second inhibitory function is lost because S-phase is now the legitimate event for a G1 cell.

The manner by which the cdc2p-cdc13p CDK inhibits re-initiation of DNA replication is not known, but it is of interest that cdc18p has cdc2p consensus phosphorylation sites that appear to be inhibitory for cdc18p function. Perhaps phosphorylation of cdc18p and of other proteins may prevent replicative complexes from being formed. This can be incorporated into a two-step mechanism for the initiation of DNA replication (Stern and Nurse, 1996). In the first step, cdc2p CDK protein kinase activity must be reduced to a level that allows formation of the initiating replicative complexes on origin regions of the DNA. In the second step, cdc2p CDK (usually complexed with cig2p) protein kinase activity must rise to a level that brings about initiation. In such a scheme the second step automatically inhibits the first step. This would prevent the re-establishment of replicative complexes on regions of the DNA that have already been replicated, thus ensuring that no region is replicated twice during one S-phase.

VIII. Evolutionary Aspects

From the above account it can be seen that the cdc2p CDK plays a key role in maintaining an orderly progression through the cell cycle. In early G1, cdc2p CDK activity is very low, allowing step one for the initiation of DNA replication. A rise of cdc2p-cig2p CDK activity at the end of G1 carries out step two and brings about S-phase, while at the same time it prevents the re-replication of any region of DNA that has already been replicated. The continued presence of cdc2p CDK activity in the form of cdc2p-cdc13p during G2 prevents another S-phase from taking place. Finally, further activation to a higher level of cdc2p-cdc13p leads to mitotic onset. To exit mitosis and enter G1 of the next cell cycle, cdc2p CDK activity must fall to a very low level again. Therefore, orderly progression through the cell cycle is driven by increasing activity associated with the cdc2p protein kinase.

Although in fission yeast these transitions are normally brought about by different cyclin complexes, primarily cdc2p-cig2p and cdc2p-cdc13p, it is possible for cell cycle progression to be regulated entirely by the single CDK cdc2p-cdc13p. This may echo the situation present in more primitive ancient eukaryotes when the cell cycle may have been controlled by a single CDK, the activity of which gradually increased throughout the cell cycle, bringing about the major cell cycle transitions and establishing the major cell cycle checkpoints.

What is not clear, however, is why the eukaryotic cell should have invested so much control in a single CDK that regulates the onset of such different events as S-phase and mitosis. It is possible that in the primitive ancient eukaryotic cell, DNA replication and chromosome segregation were not distinct events (Nurse, 1994). Bi-directional replication results in two replicating forks moving away from each other. This separating process may have also been the primitive mechanism for chromosomal segregation if there was only one origin per chromosome. The replicative complexes present at the two forks could segregate the helix containing the Watson strand away from the other helix containing the Crick strand. Thus, initiation of replication (S) and segregation (M) could have been the same process, which was controlled by the same mechanism involving a primitive CDK. As eukaryotic cells became more complex with larger amounts of DNA, they required multiple origins for replication and chromosome condensation for mitosis. As a consequence, the two events S and M had to become distinct. However, the same CDK would still be initiating both events, with low levels initiating S and higher levels M. Such evolutionary scenarios are obviously speculative but provide at least a tenable explanation for why CDKs have such a central role to play in the regulation of such different events as S-phase and mitosis during the cell cycle.

REFERENCES

Correa-Bordes, J., and Nurse, P. (1995). *Cell* **83,** 1001–1009.

D'Urso, G., Grallert, B., and Nurse, P. (1995). *J. Cell Sci.* **108,** 3109–3118.

Enoch, T., and Nurse, P. (1990). *Cell* **60,** 665–673.

Enoch, T., and Nurse, P. (1991). *Cell* **65,** 921–923.

Enoch, T., Carr, A.M., and Nurse, P. (1992). *Genes Dev.* **6,** 2035–2046.

Enoch, T., Matthias, P., Nurse, P., and Nigg, E. (1991). *J Cell Biol.* **122,** 797–807.

Fisher, D.L., and Nurse, P. (1996). *EMBO J.* **15,** 850–860.

Forsburg, S., and Nurse, P. (1991). *Nature* **351,** 245–248.

Forsberg, S., and Nurse, P. (1994). *J. Cell Sci.* **107,** 601–613.

Gould, K., and Nurse, P. (1989). *Nature* **342,** 39–45.

Gould, K., Moreno, S., Tonks, N., and Nurse, P. (1990). *Science* **250,** 1673–1576.

Gould, K., Moreno, S., Owen, D.J., Sazer, S., and Nurse, P. (1991). *EMBO J.* **10,** 3297–3309.

Grallert, B., and Nurse, P. (1996). *Genes Dev.* **10,** 2644–2654.

Hagan, I., Hayles, J., and Nurse, P. (1988). *J. Cell Sci.* **91,** 587–595.

Hayles, J., Fisher, D., Woollard, A., and Nurse, P. (1994). *Cell* **78,** 813–822.

Kelly, T.J., Martin, G.S., Forsburg, S.L., Stephen, R.J., Russo, A., and Nurse, P. (1993). *Cell* **74,** 371–382.

Moreno, S., and Nurse, P. (1990). *Cell* **61,** 549–551.

Moreno, S., and Nurse, P. (1994). *Nature* **367,** 236–242.

Moreno, S., Hayles, J., and Nurse, P. (1989). *Cell* **58,** 361–372.

Nishitani, H., and Nurse, P. (1995). *Cell* **83,** 397–405.

Nurse, P. (1975). *Nature* **256,** 547–551.

Nurse, P. (1990). *Nature* **344,** 503–508.

Nurse, P. (1994). *Cell* **79,** 547–550.

Nurse, P., Thuriaux, P., and Nasmyth, K. (1976). *Molec. Gen. Genet.* **146,** 167–178.

Russell, P., and Nurse, P. (1986). *Cell* **45,** 145–153.

Russell, P., and Nurse, P. (1987). *Cell* **49,** 559–567.

Simanis, V., and Nurse, P. (1986). *Cell* **45,** 261–268.

Stern, B., and Nurse, P. (1996). *Trends Genet.* **12,** 345–350.

ENDOGENOUS INHIBITORS OF ANGIOGENESIS

JUDAH FOLKMAN

Department of Surgery, Children's Hospital, and Departments of Surgery and Cell Biology, Harvard Medical School, Boston, Massachusetts

I. INTRODUCTION

The field of angiogenesis research began in the late 1960s from an attempt to delineate the role of neovascularization in tumor growth. Simple experiments demonstrated a severe restriction of tumor growth in isolated perfused organs in which blood vessel growth was lacking (Folkman et al., 1963). When these results were taken together with observations of retinoblastoma metastases in the human vitreous, which were similarly restricted in size to less than 1 mm^3, a hypothesis was developed in 1971 that tumors were angiogenesis-dependent and that antiangiogenic therapy could potentially be employed in cancer (Folkman, 1971, 1972).

Over the past 30 years, angiogenesis research has become a very active and broad field in biomedicine. The 1970s were devoted mainly to the development of methodology to study angiogenesis (Folkman et al., 1971; Auerbach et al., 1974; Langer and Folkman, 1976), including the first *in vitro* culture of vascular endothelial cells (Gimbrone et al., 1973, 1974a; Haundenschild et al., 1975; Haudenschild et al., 1976; Folkman and Moscona, 1978; Kadish et al., 1979; Folkman et al., 1979; Folkman and Haudenschild, 1980). Indirect evidence that tumors are angiogenesis-dependent was elucidated (Gimbrone et al., 1972, 1974b; Cavallo et al., 1972; Auerbach et al., 1975; Folkman and Klasgsbrun, 1987; Folkman, 1990). The 1980s saw the discovery of the first molecules that mediate angiogenesis, including basic fibroblast growth factor (bFGF) (Shing et al., 1984) and vascular endothelial growth factor VEGF (Senger et al., 1983; Ferrara and Henzel, 1989; Folkman et al., 1988; for review, see Folkman 1997a). Currently, there are at least 14 proteins that induce angiogenesis. Also during this decade the first inhibitors of angiogenesis were discovered (Taylor and Folkman, 1982;

The Harvey Lectures, Series 92, pages 65–82

Crum et al., 1985; Ingber et al., 1990; Kusaka et al., 1994). The role of the switch to the angiogenic phenotype during tumorigenesis was delineated (Folkman et al., 1989). It was proposed that this switch was the result of a net balance of positive and negative regulators of angiogenesis (Rastinejad et al., 1989; Bouck, 1990). The first direct evidence that tumors are angiogenesis-dependent was also reported during this time (Hori et al., 1991; Kim et al., 1993; Millauer et al., 1994). During the 1990s, the angiogenic process began to be studied by researchers in many other fields beyond cancer biology and ophthalmology. The molecular biology of vascular development progressed rapidly (Sato et al., 1995; Carmeliet et al., 1996a, 1996b; Suri et al., 1996; Eichmann et al., 1996; Vikula et al., 1996; Breier and Risau, 1996). The angiogenic process was studied in other disciplines including wound repair, reproductive biology, cardiology, and dermatology. Studies of the genetics of angiogenesis began (Carmeliet et al., 1996a, 1996b; Risau and Flamme, 1995; Ilopoulos et al.,1996; Gnarra et al., 1996; Rak et al., 1996; Olofsson et al., 1996). Investigators undertook studies of the molecular basis of endothelial cell signal transduction, (Waltenberger et al., 1994; Hughes et al., 1997; Liu et al., 1997); the molecular interactions of endothelial cells with extracellular matrix (Sage and Bornstein,1995; Madri and Williams, 1983; Pepper et al., 1996; Brooks et al., 1995; Senger et al., 1994; Nicosia and Ottinett, 1990) and with other host cells (Beck and D'Amore, 1997); and the molecular effects of tumor angiogenesis *per se* (Boucher et al., 1996; Melder et al., 1996; Vazquez, 1997). Potent angiogenesis inhibitors were discovered as internal fragments of larger proteins and were found to be specific inhibitors of endothelial cell proliferation (O'Reilly et al., 1994, 1996, 1997). The concept that tumor burden could regulate angiogenesis locally and at distant sites was introduced (O'Reilly et al., 1994). Antiangiogenic therapy was proposed as a strategy to by-pass drug resistance (Kerbel, 1991) and was then demonstrated experimentally (Boehm et al., 1997). Blocked angiogenesis was shown to be a cause of prolonged tumor dormancy (Holmgren et al., 1995). Clinical trials of angiogenesis inhibitors began (Folkman, 1995a, 1995b). Angiogenesis was found to be a prognostic indicator for a wide variety of human tumors (Weidner et al., 1991, 1993; Gasparini, 1996).

This brief summary does not do justice to the breadth of this field, and space does not allow an inclusive review of the many outstanding reports by researchers in it.

In this paper I focus on tumor angiogenesis and specifically on new concepts of cancer biology that have emerged as a result of the discovery of endogenous inhibitors of angiogenesis.

II. DISCOVERY OF ENDOSTATIN AND ANGIOSTATIN

The discovery of angiostatin resulted from our attempt to uncover the mechanism of suppression of metastatic growth by a primary tumor. It had long been known that removal of certain primary tumors in animals and in patients was sometimes followed by rapid growth of metastases (see Gorelik, 1983; Himmele et al., 1986). In 1993 Prehn reviewed the evidence that tumor mass could suppress tumor growth both locally and at remote sites, but pointed out the lack of a satisfactory mechanism to explain this. In a separate series of experiments we had analyzed the growth of fragmented spleens in the peritoneal cavity of splenectomized mice, a phenomenon known to surgeons as splenosis. Implanted spleen fragments, as small as clumps of cells, grew to form minute spleens with normal vascularization and histology, but the total mass did not exceed the weight of the original spleen (Saenz and Folkman, 1990). This indicated that even normal organ growth was governed in some unknown manner by total mass. In a parallel effort, Douglas Hanahan and I were investigating the switch to the angiogenic phenotype of spontaneous tumors in transgenic mice (Folkman et al., 1989; Hanahan and Folkman, 1996). We were puzzled by the co-existence of high levels of angiogenic stimulators and inhibitors that were found in the tumor at the time of the switch. Then in 1989, Noel Bouck reported that the switch to angiogenesis during tumorigenesis of transformed hamster cells was associated with downregulation of an inhibitor of angiogenesis, thrombospondin (Rastinejad et al., 1989) She suggested that the switch to the angiogenic phenotype could be the result of a shift in the net balance of positive and negative regulators of angiogenesis (Bouck, 1990).

Taken together, all of these findings suggested to me that a primary tumor might suppress growth of its distant metastases by releasing an angiogenesis inhibitor into the circulation. According to this hypothesis, the primary tumor would stimulate angiogenesis only in its own tumor bed, for example by local production of bFGF or VEGF, both of which have short half-lives in the circulation, three and 30 minutes, respectively. Therefore, if a putative endothelial inhibitor, perhaps having been downregulated during tumorigenesis, had a longer half-life in the circulation, then inhibitory activity would gradually accumulate in the circulation. When a metastasis began to grow, its own balance of angiogenesis stimulators and inhibitors would be overcome by the addition of circulating inhibitor from the primary. Therefore, in experimental animals, the first tumor to reach a sufficient size to release an angiogenesis inhibitor into the circulation would suppress a secondary tumor. This hypothesis predicts that if tumor angiogenesis depends upon an alteration in the net balance between angiogenesis stimulators and inhibitors produced by the tumor, then a small difference in the time necessary to clear these molecules from the circulation could permit a tumor to *stimulate angiogenesis in its own vascular bed, but inhibit angiogenesis in the vascular bed of a remote metastasis.*

Michael O'Reilly came to my laboratory as a postdoctoral fellow in July 1991 and began to test this hypothesis by screening a variety of transplantable murine tumors for their ability to suppress metastases. A Lewis lung carcinoma was the most efficient. A subclone was isolated that could not suppress metastasis. When the metastasis-suppressing primary tumor was present in the dorsal subcutaneous position, microscopic lung metastases remained dormant at a diameter of less than 200 μm surrounding a preexisting microvessel, but revealed no new vessels (by staining with antibody to von Willebrand factor). Within five days after surgical removal of the primary tumor, lung metastases became highly angiogenic and grew rapidly, killing their host by 15 days (O'Reilly et al., 1994). This striking evidence that the *primary tumor could suppress angiogenesis in its secondary metastases* by a circulating inhibitor was further supported by the demonstration that a primary tumor could also suppress corneal angiogenesis induced by an implanted pellet of bFGF. O'Reilly then succeeded in purifying this inhibitor from the serum and urine of tumor-bearing animals, a Herculean effort that took three years. It was a 38-kD internal fragment identical in amino acid sequence to the first four kringle structures of plasminogen, and it was named angiostatin. Angiostatin specifically inhibited the proliferation of growing vascular endothelial cells and had no effect on resting confluent endothelial cells or on other cell types, including smooth muscle cells, epithelial cells, fibroblasts, and tumor cells. It inhibited angiogenesis in the chick embryo, and when administered systemically to mice bearing Lewis lung carcinomas, inhibited growth of lung metastases after removal of the primary tumor. It also inhibited growth of primary tumors by up to 98% (O'Reilly et al., 1996) and was able to induce regression of large tumors (1–2% of body weight) and maintain them at a microscopic dormant size. Angiostatin had no effect on blood coagulation; plasminogen had no antiangiogenic effect. The half-life of angiostatin in the circulation was approximately 2.5 days, consistent with the prediction of the original hypothesis. However, the prediction that the inhibitor itself would originate from the tumor cells was not substantiated. Instead, M. Moses and O'Reilly in our laboratory found that the Lewis lung cells released a metalloproteinase *in vitro* that could generate angiostatin from plasminogen (unpublished personal communication; Moses, 1997). Others found that a 72-kD gelatinase could cleave biologically active angiostatin from plasminogen (Stetler-Stevenson and Corcoran, 1997) or that the Lewis lung carcinoma cells stimulated macrophages to enzymatically release angiostatin from plasminogen (Dong et al., 1997). Angiostatin was also cleaved from plasminogen by certain sulfhydryl donors present on the cell surface (Stathakis et al., 1997). When plasmin was co-incubated with N-acetylcysteine (Gately et al., 1997), autoproteolysis led to production of angiostatin.

Based upon the same rationale and strategy, O'Reilly isolated and purified another angiogenesis inhibitor from a murine hemangioendothelioma. This inhibitor, called endostatin (O'Reilly et al., 1997), is a 20-kD protein with an N-terminal amino acid sequence identical to the carboxyterminus of collagen XVIII (Oh et al., 1994; Rhen and Pihlajaniemi, 1994). It was purified directly from tumor cell-conditioned medium. Endostatin is also a specific inhibitor of endothelial proliferation and has no effect on resting endothelial cells or on other cell types. It is slightly more potent than angiostatin, and also causes regression of large tumors to a microscopic size.

III. A Mechanism for Tumor Dormancy — Blocked Angiogenesis

When angiostatin or endostatin were administered systemically to tumor-bearing mice, tumor cell apoptosis increased (e.g., from 2% to 7–8%) as neovascularization decreased, but the rate of tumor cell proliferation was the same in the angiogenic as in the non-angiogenic tumors (e.g., approximately 38–40%) (Holmgren et al., 1995; O'Reilly et al., 1996, 1997) (Fig. 1). In other words, angiogenesis is inversely correlated with tumor cell apoptosis but is independent of tumor cell proliferation. This pattern was also observed in dormant metastases in which angiogenesis was suppressed by a primary tumor (O'Reilly et al., 1994). A new mechanism of prolonged tumor dormancy due to blocked angiogenesis in which tumor cell apoptosis and proliferation are balanced has therefore emerged from the study of these powerful endogenous angiogenesis inhibitors. Because of the persistently high tumor cell proliferation rate, tumors held dormant by antiangiogenic therapy may be very sensitive to short-term conventional chemotherapy.

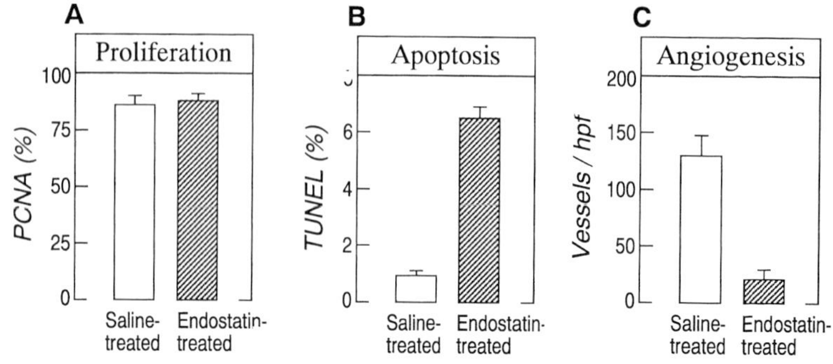

Fig 1. Endostatin therapy inhibits angiogenesis, decreases microvessel density, and causes regression of Lewis lung carcinomas in mice. This antiangiogenic therapy results in inhibition of angiogenesis and an increase in apoptosis in Lewis lung carcinoma, but does not decrease tumor cell proliferation. (From O'Reilly et al., 1997, with permission from *Cell*).

IV. The "Indolent Tumor" — A Proposed Angiogenic Mechanism

Although the conventional wisdom is that indolent, slowly growing human tumors have a significantly lower growth fraction than rapidly growing tumors, there are many exceptions. In our laboratory Yihai Cao transfected the cDNA for mouse angiostatin into murine fibrosarcoma cells (Cao et al., 1998). Transfected tumor cells that were either high or low producers of angiostatin were cloned and implanted subcutaneously in mice. The resulting tumors grew very slowly if they were high producers of angiostatin, and rapidly if they were low producers (Fig. 2). There was, however, no significant difference in the growth fraction of the two tumor cell populations.

A human counterpart to this experiment may be prostate carcinomas confined to the prostate where the growth fraction is low, 1.54% of tumor cells proliferating per day (Berges et al., 1995). The doubling time of the tumor mass is also slow, 577 ± 68 days. In contrast, in rapidly growing bone metastases of this tumor in which the doubling time of tumor mass is increased by 10-fold to 54 ± 5 days, the growth fraction is not significantly different from the indolent tumor, i.e., 2.04% of tumor cells proliferating per day. Microvessel density was not determined in this study; indolent human prostate cancer may become a model of the control of tumor mass by the intensity of angiogenesis. Therefore, a tumor with a high growth fraction may

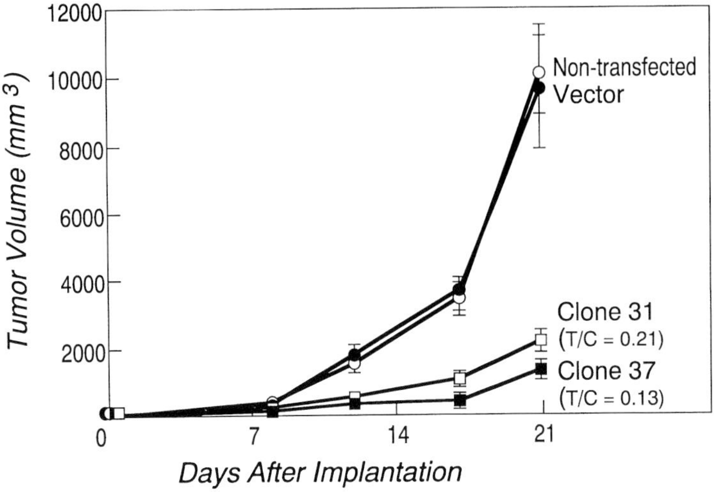

Fig. 2. Angiostatin cDNA is transfected into murine T241 fibrosarcoma cells, and high and low producers are selected from cells that have been cloned *in vitro*. When these tumor cells are inoculated subcutaneously into mice, the high producers of angiostatin yield slowly growing indolent tumors. The *in vitro* proliferation rates of the high and low producers of angiostatin are not significantly different. (From Cao et al., with permission from *J. Clin. Invest.*)

remain indolent if its angiogenic activity is low or blocked completely. Conversely, a tumor with a low growth fraction may expand either slowly or rapidly depending upon its angiogenic intensity.

V. Antiangiogenic Therapy as a Strategy to By-Pass Drug Resistance

Robert Kerbel hypothesized in 1991 that inhibition of angiogenesis could be employed as a strategy to by-pass drug resistance (Kerbel, 1991). We now have experimental evidence for this concept. Mice bearing either melanoma, fibrosarcoma, or Lewis lung carcinoma were treated with endostatin until the large tumors (1–2% of body weight) regressed completely by gross examination. Treatment was then discontinued and resumed when the tumors had regrown to the same size as before the first treatment. No drug resistance developed despite repeated cycles of therapy over periods of 80 to 185 days (Boehm et al., 1997) (Fig. 3).

An unexpected finding during this experiment was that for each tumor type there was a given time period of endostatin therapy beyond which there was no tumor regrowth after discontinuation of therapy. Tumors remained dormant at a microscopic or barely visible size while the animals were observed for an additional 103 to 165 days after the beginning of the dormant state. The microscopic residual tumors were viable by histology and ranged in size from 0.06 mm^3 to 2.23 mm^3. They were avascular and showed the pattern of dormant tumors with blocked angiogenesis, i.e., high tumor cell proliferation balanced by a high apoptosis rate. When they were transplanted to a new location in the host mice, rapidly growing tumors appeared at the new transplantation site and at the old wound site. This indicates that *exogenous* angiogenesis induced by a wound, can override the self-sustained dormancy resulting from inhibition of *endogenous* angiogenesis. The long-term self-sustained dormancy in these lesions was not due to an immune response, because subcutaneous implantation of the same type of tumor cells from cell culture led to rapidly growing, well-vascularized tumors. At this writing, the mechanism of the "self-sustained dormancy" after discontinuation of therapy remains unknown. However, it is consistent with dormancy due to blocked angiogenesis. One can speculate that it could be due in part to senescence of endothelial cells in the vascular bed of the tumor, or to accumulation of extracellular matrix that is inhibitory to endothelial growth, or to selection of tumor cells that lack angiogenic activity. The latter hypothesis is of interest because experimental studies reveal that even in a well-vascularized tumor only a subpopulation of tumor cells is angiogenic.

Fig. 3. Cycled endostatin therapy of tumors grown in the subcutaneous dorsum of mice to demonstrate lack of drug resistance to antiangiogenic therapy. **a:** Lewis lung carcinoma. When tumor volume reached 250 to 400 mm^3, endostatin therapy was initiated. Tumors in four untreated animals grew to 10,005 ± 131 mm^3 by day 27 (data not graphed). **b:** T241 fibrosarcoma. Endostatin therapy was initiated when the mean tumor volume was 200 mm^3 and in subsequent cycles when the tumor volume was greater than 300 mm^3. Tumors in four untreated animals grew to 9127 ± 1183 mm^3 by day 28. **c:** B16F10 melanoma (a subclone of cells supplied by Isaiah Fidler, Houston). Endostatin therapy was initiated when the mean tumor volume was 230 mm^3 for the initial cycle and at 350 mm^3 for the second cycle. Tumors in four untreated animals grew to 7431 ± 1122 mm^3 by day 41. (From Boehm et al., 1997, with permission from *Nature*.)

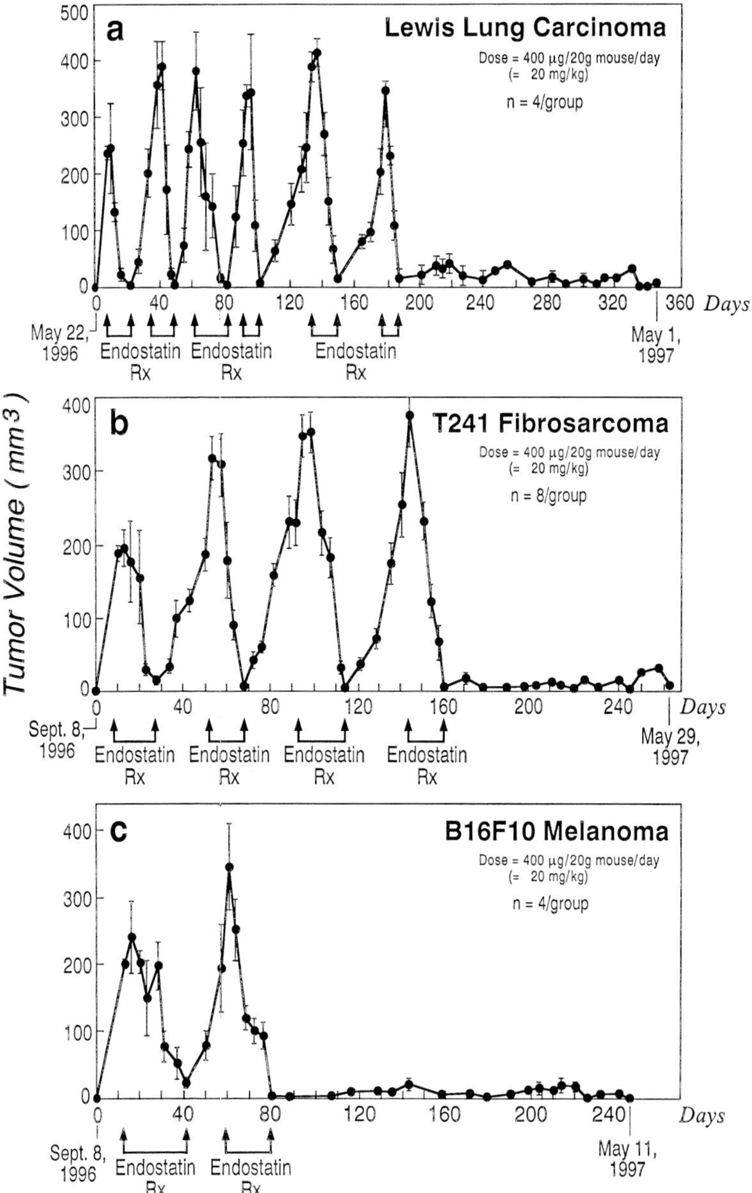
a
Lewis Lung Carcinoma
Dose = 400 μg/20g mouse/day
(= 20 mg/kg)
n = 4/group
May 22, 1996
Endostatin Rx
Endostatin Rx
Endostatin Rx
May 1, 1997
Days
Tumor Volume (mm3)
b
T241 Fibrosarcoma
Dose = 400 μg/20g mouse/day
(= 20 mg/kg)
n = 8/group
Sept. 8, 1996
Endostatin Rx
Endostatin Rx
Endostatin Rx
Endostatin Rx
May 29, 1997
Days
c
B16F10 Melanoma
Dose = 400 μg/20g mouse/day
(= 20 mg/kg)
n = 4/group
Sept. 8, 1996
Endostatin Rx
Endostatin Rx
May 11, 1997
Days

Figure 3.

These cells appear to induce the neovasculature upon which the non-angiogenic cells thrive. In an analysis of well-vascularized transplantable tumors (i.e., sarcoma 180) usually less than 25% of the tumor cells were mitogenic for endothelial cells (Folkman, unpublished data). This was determined by incubating capillary endothelial cells in conditioned media from tumor cells cloned by dilution from vascularized tumors. Tumor cells that were not mitogenic for endothelium did not form vascularized tumors upon subcutaneous injection into mice, but remained as microscopic dormant tumors. When they were mixed with tumor cells that were mitogenic for endothelium, the percentage of vascularized tumors was proportional to the tumor cells that were mitogenic for endothelium. These data suggest that while angiogenic and non-angiogenic tumor cells would be difficult to distinguish within a primary tumor, after they have metastasized, only the angiogenic tumor cells would be expected to become grossly visible metastases. The non-angiogenic tumor cells would remain dormant at a microscopic size (Holmgren et al., 1995).

In contrast to antiangiogenic therapy, conventional cytotoxic therapy commonly induces drug resistance. Tumor cells are highly heterogenous and have a high mutation rate and are thus capable of generating resistant clones. Endothelial cells are relatively homogenous and do not mutate. Thus, like bone marrow, they do not usually develop drug resistance.

The lack of resistance to an angiogenesis inhibitor would only be expected to apply to inhibitors such as angiostatin, endostatin, or TNP-470 (synthetic analogue of fumagillin) (Ingber et al., 1990). An angiogenesis inhibitor such as interferon-alpha, which acts by blocking synthesis and protein production of the angiogenic factor bFGF from tumor cells (Singh et al., 1995) might eventually induce drug resistance if tumor cells appeared that were dependent upon an angiogenic factor other than bFGF (Fig. 4). However, the occurrence of drug resistance could also depend on the mutation rate and thus the degree of malignancy of the tumor. For example, we have not observed the development of drug resistance in infants treated for more than one year with interferon-alpha for life-threatening hemangioma (Folkman et al., 1997), or in a child with a low-grade giant cell tumor of the bone treated for one year. In contrast, bladder cancer does develop resistance to interferon-alpha. Similarly, tumors could eventually develop drug resistance to antibodies directed against specific angiogenic factors (e.g., bFGF or VEGF) if the tumors began to generate cells that were angiogenic by production of a different angiogenic factor. Therefore, an anti-VEGF antibody might be more useful for short-term anti-cancer therapy, or in com-

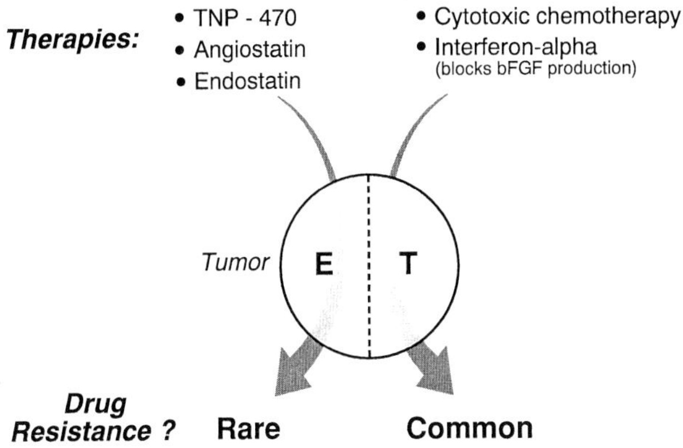

Fig. 4. Diagram of lack of drug resistance by therapy of the endothelial cell compartment of a tumor vs the tumor cell compartment. Within a tumor, the tumor cell population (T) commonly develops resistance to a wide variety of cytotoxic drugs, whereas the endothelial cell population (E) does not. Thus, therapies directed against endothelial cells, rather than tumor cells, are less likely to induce drug resistance.

bination with other agents, or in non-neoplastic disease such as ophthalmologic neovascularization, where the likelihood of resistance due to cell mutation is low.

One can speculate that if and when the first angiogenesis inhibitor that does not induce drug resistance is approved for general medical use, it could change the practice of oncology because tumor markers would assume increasing importance as guides to therapy. For example, if a patient with prostate cancer had a rising prostate specific antigen in his serum (PSA) after completion of therapy, it might only be necessary to resume antiangiogenic therapy until the PSA returned to normal.

In our experiments demonstrating lack of drug resistance by endostatin, we found that when angiostatin and endostatin were administered together (each at 20 mg/kg/day, e.g., 400 micrograms/mouse/day) for only 25 days, Lewis lung tumors of 300 mm^3 (1.5% of body weight) regressed completely and the animals were cured i.e., no tumor recurrence by one year off therapy and no detectable tumor by histological analysis (O'Reilly et al., unpublished data).

In summary, repetitive cycles of endostatin therapy alone induced a form of self-sustained dormancy, angiostatin and endostatin together were curative, and neither antiangiogenic regimen induced drug resistance.

V. A Two-Cell Compartment Model of Tumor Growth

A. *Tumor cell-capillary units*

In order to understand how inhibition of capillary blood vessel growth can bring about tumor regression in experimental animals, it is necessary to know how tumor cells are associated with microvascular endothelial cells in a three-dimensional configuration. Once angiogenesis has been initiated, capillary sprouts and loops converge upon a local tumor that may be no larger than 1 mm^3 or less. The sprouts lack smooth muscle. Tumor cells form perivascular cuffs of approximately 4–5 cell layers radius, but this can vary from one tumor to another. This can be observed in confocal microscopic reconstructions of human breast cancer (Folkman, 1997a). The important point is the difference from normal tissue, where most parenchymal cells, such as in the liver, form only a single cell layer around each capillary. A tumor of 1 gram contains approximately 10^8 to 10^9 tumor cells and approximately 2×10^7 endothelial cells (Modzelewski et al., 1994). Therefore, approximately 5 to 50 or more tumor cells may be supported by the length of capillary vessel occupied by a single endothelial cell. When the growth of this endothelial cell is arrested, the expansion of the cylindrical mass of tumor cells surrounding this vascular segment will also be restricted. When this endothelial cell undergoes apoptosis, the surrounding microcylinder of tumor cells may die. A recent demonstration of this model was shown in animal tumors that secreted VEGF under the control of a tetracycline promoter (Benjamin and Keshet, 1997). Suppression of VEGF production led to endothelial apoptosis and death of surrounding tumor cells.

New tumor vessels do more than just perfuse a tumor with blood which contains nutrients and oxygen and which removes catabolites. In addition to this "perfusion" effect, vascular endothelial cells also produce "paracrine" factors that can stimulate tumor cell proliferation and motility. These endothelial-derived paracrine factors include bFGF, platelet-derived growth factor (PDGF-BB) and insulin-like growth factor-1 (IGF-1), among a total of at least 20 such factors (Rak, 1996). This paracrine effect can also be observed *in vitro* where tumor cells preferentially grow apposed to capillary endothelial tubes despite the absence of blood flow (Nicosia et al., 1984). If antiangiogenic therapy arrests the growth of the intratumoral endothelial cell mass or causes this cell population to regress, this could result in the disappearance of growth factors and/or survival factors. In animal tumors (Holmgren et al., 1995; O'Reilly et al., 1994, 1996, 1997) and in human tumors (Caide and Tanigawa, 1997), the induction of angiogenesis correlates with at least a sevenfold decrease in apoptosis. In contrast, the proliferation rate of tumor cells remains equally high in the prevascular as well as in the vascularized tumor (up to 40% by BrdU). This pattern is observed in a wide range of different tumor types. It indicates that tumor expansion after neovascularization may depend mainly on restriction of tumor cell apoptosis. Analogous systems are found in *myc*-dependent lymphoma cells or in fibroblasts with dysregulated *myc*-expression. Both cell types grow in the presence of IGF-1 or IGF-2, respectively, but die when these factors are withdrawn (Evans, 1994). It is not clear whether reduction of tumor growth by blocked angiogenesis may also depend on a mechanism by which endothelial cells are activated to directly lyse tumor cells (Li et al., 1991). Also, it remains to be demonstrated whether angiostatin or endostatin therapy can suppress endothelial production of paracrine growth or survival factors for tumor cells.

B. Endothelial and Tumor Cell Compartments

The interaction between endothelial cells in microvessels and the tumor cells that surround these vessels is bidirectional. The tumor cells drive endothelial cells by releasing angiogenic proteins, and the endothelial cells provide paracrine stimulation of tumor cells. This working model of two interacting cell populations provides a framework to think about cancer biology in novel ways. The analysis of drug resistance in terms of endothelial cells as a target in contrast to tumor cells as a target is one example. Another example is the use of this model to think about an improvement of conventional chemotherapy. If endothelial cells can control tumor cell growth and they do not develop drug resistance, why should a tumor ever become drug-resistant to conventional cytotoxic therapy? Many cytotoxic drugs do not seem to affect angiogenesis (Steiner, 1992). Others may not inhibit endothelial cell migration. However, certain cytotoxic drugs such as cyclophosphamide are also antiangiogenic in experimental animals (Timothy Browder and also Robert D'Amato, personal communication). Timothy Browder in my laboratory has shown that cyclophosphamide therapy can be *optimized* for the endothelial cell compartment by changing its dose and schedule. This results in cure of formerly incurable Lewis lung carcinomas that were not drug-resistant. Optimization of cytotoxic therapy for the endothelial cell population can suppress the growth of drug-resistant Lewis lung carcinomas that formerly escaped cyclophosphamide therapy (Browder, unpublished data). Therefore, when the tumor cell compartment becomes drug-resistant, it may be analogous to the tumor cell that is completely refractory to endostatin from the first day of endostatin therapy.

VI. Advantages of Antiangiogenic Therapy Compared to Cytotoxic Chemotherapy for Cancer

Proof of principle that tumors are angiogenesis-dependent has been established in experimental animals, and our data reveals that antiangiogenic therapy has none of the toxic side effects or the drug resistance-inducing properties of conventional chemotherapy. This difference is due in part to the fact that a typical tumor has a small fraction of highly proliferating cells and a substantial fraction of non-growing or slowly growing cells. Conventional cytotoxic chemotherapy targets mainly proliferating cells in the tumor, but its use is limited by coincident injury to normal tissues, in accordance with the individual proliferation rates of those tissues. Repeated application of chemotherapy is necessary because the quiescent subpopulation of tumor cells may enter the cell cycle as the cells occupy the niche created by cytotoxic killing of the original growth fraction of tumor cells. Repeated cycles of chemotherapy also result in drug resistance mainly because of the genetic instability and high mutation rate of tumor cells.

In contrast, antiangiogenic therapy, which specifically targets highly proliferating *endothelial* cells in the tumor bed, restricts growth of all tumor cell populations (i.e., growing and quiescent subpopulations). Resistance does not develop because of the genetic stability and low mutation rate of endothelial cells. Moreover, the wide difference in turnover times between the tumor-associated endothelial cells (approximately five days) and the non-proliferating endothelial cells in the rest of the host (>1000 days) assures the apparent insensitivity of the normal endothelium to these novel antiangiogenic agents. Therefore, antiangiogenic therapy directed toward the tumor endothelial cell compartment circumvents two major therapeutic issues: toxicity to normal tissues and drug resistance.

VII. Future Application of Antiangiogenic Therapy in Human Cancer

It is premature to predict how these experimental results will translate to clinical application. Once powerful angiogenesis inhibitors such as angiostatin or endostatin are approved for use in cancer patients, these agents may at first be employed together with conventional chemotherapy or radiotherapy. Subsequently, they may be used after the completion of chemotherapy to maintain prolonged dormancy or to prevent the recurrence of metastases. Antiangiogenic therapy may also be used as a platform to enhance anti-cancer therapies that emerge in the future, such as immunotherapy, telomerase inhibitor therapy, or gene therapy.

Our animal experiments argue that when potent inhibitors such as angiostatin and endostatin are administered, antiangiogenic therapy is superior to conventional cytotoxic chemotherapy. Antiangiogenic therapy can regress large tumors, can maintain their dormancy, and can be curative, yet without the toxicity and drug resistance that accompany virtually all chemotherapy. Only time will tell whether antiangiogenic therapy will eventually enhance current chemotherapy or emerge as an alternative to it.

Acknowledgements

I thank Philip Hahnfeldt and Lynn Hlatky of the Dana Farber Cancer Institute, Boston, Mass., for their contributions to the concepts presented in Section VI. I thank Wendy Foss for organizing the references.

References

Auerbach, R., Kubai, L., Knighton, D., and Folkman, J. (1974). *Dev. Biol.* **41,** 391–394.
Auerbach, R., Arensman, R., Kubai, L., and Folkman, J. (1975). *Int. J. Cancer* **15,** 241–245.
Beck, L., Jr., and D'Amore, P.A. (1997). *FASEB J.* **11,** 365–373.
Benjamin, L.E., and Keshet, E. (1997). *Proc. Natl. Acad. Sci. U.S.A.* **94,** 8761–8766.
Berges, R.R., Vukanovic, J., Epstein, J.I., Carmichel, M., Cisek, L., Johnson, D.E., Veltri, R.W., Walsh, P.C., and Isaacs, J.T. (1995). *Clin. Cancer Res.* **1,** 473–480.
Bichel, P. (1971.) *Eur. J. Cancer* **7,** 349–355.
Boehm, T., Folkman, J., Browder, T., and O'Reilly, M.S. (1997). *Nature* **390,** 404–407.
Boucher, Y., Leunig, M., and Jain, R.K. (1996). *Cancer Res.* **56,** 4264–4266.
Bouck, N. (1990). *Cancer Cells-MR* **2,** 179–185.
Breier, G., and Risau, W. (1996). *Trends Cell Biol.* **6,** 454–456.
Brooks, P.C., Stromblad, S., Klemke, R., Visscher, D., Sarkar, F.H., and Cheresh, D.A. (1995). *J. Clin. Invest.* **96,** 1815–1822.
Caide I.U., and Tanigawa N. (1997). *Cancer Res.* **57,** 221–224.
Carmeliet, P., Mackman, N., Moons, L., Luther, T., Gressens, P., Van Vlaenderen, I., Demunck, H., Kasper, M., Breier, G., Evrad, P., Muller, M., Risau, W., Endington, T., and Collen, D. (1996a). *Nature* **383,** 73–75.
Carmeliet, P., Ferreira, V., Breier, G., Pollefeyt, S., Kieckens, L., Gertsenstein, M., Fahrig, M., Vandenhoeck, A., Harpal, K., Eberhardt, C., Declercq, C., Pawling, J., Moons, L., Collen, D., Risau, W., and Nagy, A. (1996b). *Nature* **380,** 435–439.
Cavallo, T., Sade, R., Folkman, J., and Cotran, R.S. (1972). *J. Cell Biol.* **54,** 408–420.
Crum, R., Szabo, S., and Folkman, J. (1985). *Science* **230,** 1375–1378.
Dong, Z., Kumar, R., Yang, X., and Fidler, I.J. (1997). *Cell* **88,** 801–810.

Eichmann, A., Marcell, C., Breant, C., and Le Douarin, N.M. (1996). *Gene* **174,** 3–8.

Evans G.I. (1994). *Trends Cell. Biol.* **4,** 191–192.

Ferrara, N., Henzel, W.J. (1989). *Biochemical and Biophysical Research Communications* **161,** 851–855.

Folkman, J. (1971). *N. Engl. J. Med.* **285,** 404–405.

Folkman, J. (1972). *Ann. Surg.* **175,** 409–416.

Folkman, J. (1990). *J. Natl. Cancer Inst.* **82,** 4–6.

Folkman, J. (1995a). *Nature Med.* **1,** 27–31.

Folkman, J. (1995b). *N. Engl. J. Med.* **333,** 1757–1763.

Folkman, J. (1997a) "Cancer: Principles and Practice of Oncology, Fifth Edition." Lippincott-Raven Publishers, Philadelphia, pp. 3075–3085.

Folkman, J. (1997b). *Nature Biotechnol.* **15,** 510.

Folkman, J., and Haudenschild, C. (1980). *Nature* **288,** 551–556.

Folkman, J., and Klagsbrun, M. (1987). *Science* **235,** 442–447.

Folkman, J., and Moscona, A. (1978). *Nature* **273,** 346–349.

Folkman, J., Long, D.M., and Becker, F.F. (1963). *Cancer* **16,** 453–467.

Folkman, J., Merler, E., Abernathy, C., and Williams, C. (1971). *J. Exp. Med.* **133,** 275–288.

Folkman, J., Haudenschild, C.C., and Zetter, B.R. (1979). *Proc. Natl. Acad. Sci. U.S.A.* **76,** 5217–4221.

Folkman, J., Klagsbrun, M., Sasse, J., Wadzinski, M., Ingber, D.E., and Vlodavsky, I. (1988). *Am. J. Pathol.* **130,** 393–400.

Folkman, J., Watson, K., Ingber, D., and Hanahan, D. (1989). *Nature* **339,** 58–61.

Folkman, J., Mulliken, J.B., and Ezekowitz, R.A.B. (1997). "The Clinical Applications of the Interferons," pp. 255–265, Chapman & Hall Medical, London.

Gasparini, G. (1996). *Eur. J. Cancer* **32A,** 2485–2493.

Gately, S., Twardowski, P., Stack, M.S., Cundiff, D.L., Grella, D., Castellino, R.J., Enghild, J., Kwaan, H.C., Lee, F., Kramer, R.A., Volpert, O., Bouck, N., and Soff, G.A. (1997). *Proc. Natl. Acad. Sci.* **94,** 10868–10872.

Gimbrone, M.A., Jr., Leapman, S., Cotran, R.S., and Folkman, J. (1972). *J. Exp. Med.* **136,** 261–276.

Gimbrone, M.A., Jr., Cotran, R.S., and Folkman, J. (1973). *Ser. Haematol* **4,** 453–459.

Gimbrone, M.A., Jr., Cotran, R.S., and Folkman, J. (1974a). *J. Cell Biol.* **60,** 673–684.

Gimbrone, M.A., Jr., Cotran, R.S., and Folkman, J. (1974b). *J. Natl. Cancer Inst.* **52,** 423–427.

Gnarra, J.R., Zhou, S., Merrill, M.J., Wagner, J.R., Krumm, A., Papavassiliou, E., Oldfield, E.H., Klausner, R.D., and Linehan, V.M. (1996). *Proc. Natl. Acad. Sci. U.S.A.* **93,** 10589–10594.

Gorelik, E. (1983). *Adv. Cancer Res.* **39,** 71–120.

Hanahan, D., and Folkman, J. (1996). *Cell* **86,** 353–364.

Haudenschild, C.C., Cotran, R.S., Gimbrone, M.A., and Folkman, J. (1975). *J. Ultrastruct. Res.* **50,** 22–32.

Haudenschild, C.C., Zahniser, D., Folkman, J., and Klagsbrun, M. (1976). *Exp. Cell. Res.* **98,** 175–183.

Himmele, J.C., Rabenhorst, B., and Werner, D. (1986). *J. Cancer Res. Clin. Oncol.* **111,** 160–165.

Holmgren, L., O'Reilly, M.S., and Folkman, J. (1995). *Nature Med.* **1,** 149–153.

Hori, A., Sasada, R., Matsutani, E., Naito, K., Sakura, Y., Fijita, T., and Kozai, Y. (1991). *Cancer Res.* **51,** 6180–6184.

Hughes, P.E., Renshaw, M.W., Pfaff, M., Forsyth, J., Keivens, V.M., Schwartz, M.A., and Ginsberg, M.H. (1997). *Cell* **88,** 521–530.

Ilopoulos, O., Levy, A.P., Jiang, C., Kaelin, W.G., Jr., and Goldberg, M.A. (1996). *Proc. Natl. Acad. Sci. U.S.A.* **93,** 10595–10599.

Ingber, D., Fujita, T., Kishimoto, S., Sudo, K., Kanamaru, T., Brem, H., and Folkman, J. (1990). *Nature* **348,** 55–57.

Kadish, J.L., Butterfield, C.E., and Folkman, J. (1979). *Tiss. Cell* **11,** 99–108.

Kerbel, R.S. (1991). *BioEssays* **1,** 31–36.

Kim, K.J., Li, B., Winer, J., Armanini, M., Gillett, N., Phillips, H.S., and Ferrara, N. (1993). *Nature* **362,** 841–844.

Kusaka, M., Sudo, K., Matsutani, E., Kozai, Y., Marui, S., Fujita, T., Ingber, D., and Folkman, J. (1994). *Br. J. Cancer* **69,** 212–216.

Langer, R., and Folkman, J. (1976). *Nature* **263,** 797–880.

Li L, Nicolson G.L., and Fidler I.J. (1991). *Cancer Res.* **51,** 245–254.

Liu, J., Oh, P., Horner, T., Rogers, R.A., and Schnitzer, J.E. (1997). *J. Biol. Chem.* **272,** 7211–7222.

Madri, J., and Williams, S.K. (1983). *J. Cell Biol.* **97,** 153–165.

Melder, R.J., Koenig, G.C., Witwer, B.P., Safabakhsh, N., Munn, L.L., and Jain, R.K. (1996). *Nature Med.* **2,** 992–997.

Millauer, B., Shawyer, L.K., Plate, K.H., Risau, W., and Ullrich, A. (1994). *Nature* **367,** 576–579.

Modzelewski RA, Davies P, Watkins S.C., Auerbach R., Chang M.-J., and Johns, C.S. (1994). *Cancer Res.* **54,** 336.

Moses, M.A. (1997). *Stem Cells* **15,** 180–189.

Nicosia, R.F., and Ottinett, A. (1990). *Lab. Invest.* **63,** 115–122.

Nicosia, R., McCormick, J., and Bielunas, J. (1984). *Scan. Electron Micros.* **2,** 793–799.

Oh, S.K., Kamagata, Y., Muragaki, Y., Timmons, S., Ooshima, A., and Olsen, B.R. (1994). *Proc. Natl. Acad. Sci. U.S.A.* **91,** 4229–4233.

Olofsson, B., Pajusola, K., von Euler, G., Chilov, D., Alitalo, K., and Eriksson, U. (1996). *J. Biol. Chem.* **271,** 19310–19317.

O'Reilly, M.S., Holmgren, L., Shing, Y., Chen, C., Rosenthal, R.A., Moses, M., Lane, W.S., Cao, Y., Sage, E.H., and Folkman, J. (1994). *Cell* **79,** 315–328.

O'Reilly, M.S., Holmgren, L., Chen, C., and Folkman, J. (1996). *Nature Med.* **2,** 689–692.

O'Reilly, M.S., Boehm, T., Shing, Y., Fukai, N., Vasios, G., Lane, W.S., Flynn, E., Birkhead, J.R., Olsen, B.R., Folkman, J. (1997). *Cell* **88,** 277–285.

O'Reilly, M.A., Wiederschain, D., Stetler-Stevenson, W., Folkman, J., and Moses, M.A., personal communication, (1997).

Pepper, M.S., Montesano, R., Mandriota, S.J., Orci, L., and Vassalli, J.-D. (1996). *Enzyme Proteins* **49,** 138–162.

Prehn, R.T. (1993). *Cancer Res.* **53,** 3266–3269.

Rak, J., Filmus, J., and Kerbel, R.S. (1996). *Eur. J. Cancer* **32A**, 2438–2450.

Rastinejad, F., Polverini, P.J., and Bouck, N.P. (1989). *Cell* **56,** 345–355.

Rehn, M., and Pihlajaniemi, T. (1994). *Proc. Natl. Acad. Sci. U.S.A.* **91,** 4234–4238.

Risau, W., and Flamme, I. (1995). *Ann. Rev. Cell. Biol.* **11,** 73–91.

Saenz, N.C., and Folkman, J. (1990). *Surg. Forum* **41,** 566–569.

Sage, E.H., and Bornstein, P. (1995). "Extracellular Matrix Molecules: A Practical Approach." Oxford University Press, New York.

Sato, T.N., Tozawa, Y., Deutsch, U., Wolburg-Buchholz, K., Fujiwara, Y., Gendom-Maguire, M., Gridley, T., Wolburg, H., Risau, W., and Qin, Y. (1995). *Nature* **376,** 70–74.

Senger, D.R., Galli, S.J., Dvorak, A.M., Perruzzi, C.A., Harvey, V.S., and Dvorak, H.F. (1983). *Science* **219,** 983–985.

Senger, D.R., Brown, L.F., Claffey, K.P., and Dvorak, H.F. (1994). *Invas. Metas.* **95,** 385–394.

Shing, Y., Folkman, J., Sullivan, R., Butterfield, C., Murray, J., and Klagsbrun, M. (1984). *Science* **223,** 1296–1298.

Singh, R.K., Gutman, M., Bucana, C.D., Sanchez, R., Llansa, N., and Fidler, I.J. (1995). *Proc. Natl. Acad. Sci. U.S.A.* **92,** 4562–4566.

Stathakis, P., Fitzgerald, M., Matthias, L.J., Chesterman, C.N., and Hogg, P.J. (1997). *J. Biol. Chem.* **272,** 20641–20645.

Steiner, R. (1992) *In* "Angiogenesis Key Principles – Science – Technology – Medicine" (R. Steiner, P. Weisz, and R. Langer, eds.), pp. 449–454, Birkhauser Verlag, Basel, Switzerland.

Stetler-Stevenson, W.G., and Corcoran, M.L. (1997). *Experientia Suppl.* **79,** 413–318.

Suri, C., Jones, P.F., Patan, S., Bartunkova, S., Maisonpierre, P.C., Davis, S., Sato, T.N., and Yancopoulos, G.D. (1996). *Cell* **87,** 1171–1180.

Taylor, S., and Folkman, J. (1982). *Nature* **297,** 307–312.

Vazquez, F., Hastings, G., Lombardo, M., Oikemus, S., and Iruela-Arispe, M.L. (1997). *FASEB J.* **11,** A336.

Vikula, M., Boon, L.M., Calvert, J.T., Diamonti, A.J., Goumnerov, B., Pasyk, K.A., Marchuk, D.A., Warman, M.L., Cantley, L.C., Mulliken, J.B., and Olsen, B.R. (1996). *Cell* **87,** 1181–1190.

Waltenberger, J., Claesson-Welsh, L., Siegbahn, A., Shibuya, M., and Heldin, C.-H. (1994). *J. Biol. Chem.* **269,** 26988–26995.

Weidner, N., Semple, J.P., Welch, R., and Folkman, J. (1991). *N. Engl. J. Med.* **324,** 1–8.

Weidner, N., Carroll, P.R., Flax, J., Blumenfeld, W., and Folkman, J. (1993). *Am. J. Pathol.* **143,** 401–409.

BMPS: MULTIFUNCTIONAL REGULATORS OF MAMMALIAN EMBRYONIC DEVELOPMENT

BRIGID L.M. HOGAN

Howard Hughes Medical Institute and Department of Cell Biology, Vanderbilt University Medical Center, Nashville, Tennessee

I. INTRODUCTION

One of the most exciting advances in embryology over the past decade has been the realization that the growth and patterning of developing tissues and organs is controlled by a relatively small number of gene families that have been conserved during evolution. Indeed, it is even possible to argue that organisms such as annelid worms, insects, and vertebrates all evolved from a common segmented ancestor (known as Urbilateria), and in doing so built upon a few basic genetic mechanisms for establishing the embryonic body plan, rather than invent new ones (DeRobertis, 1997). Some of the conserved gene families encode DNA binding proteins that regulate programs of gene expression at the transcriptional level. Others encode secreted proteins that mediate the specific interactions between embryonic tissues that result in changes in cell proliferation, differentiation, or developmental fate. Families of such signaling proteins include the fibroblast growth factors (Fgfs), the hedgehogs (Hhs), proteins related to wingless and the product of the *int*-1 locus (the Wnts), and the bone morphogenetic proteins (Bmps). The latter is one of the largest families, and has been the topic of intense interest since the first members — *Drosophila* Decapentaplegic (Dpp), *Xenopus* Vg1, and mammalian Bmp2, 4 and 6 (also known as Vgr1) — were reported in the late 1980's (for extensive reviews see Harland, 1994; Hogan, 1996; Kingsley, 1994; McPherron and Lee, 1996).

II. BMP SIGNALING INVOLVES A CONSERVED SYSTEM OF POSITIVE AND NEGATIVE REGULATING FACTORS

The Bmps constitute a subgroup of the transforming growth factor β (Tgfβ) superfamily. There are currently over 20 members of this subgroup, but un-

The Harvey Lectures, Series 92, pages 83–98

fortunately the nomenclature is still complex and unstandardized, with some members also being known as growth and differentiation factors (Gdfs) or as osteogenic proteins (Ops). In addition to the ligands, it is becoming clear that there is an evolutionarily conserved system of associated factors designed to regulate Bmp activity, availability, and downstream signaling in the embryo. This system is summarized in Figure 1 and will be briefly described here. Bmps are synthesized as large precursors that dimerize and are cleaved to give a secreted, C-terminal mature protein folded into a compact three-dimensional "cysteine-knot" structure (Griffith et al., 1996). When isolated and purified, this C-terminal dimer has biological activity on target cells. However, there is now considerable evidence that, *in vivo*, the availability of at least one of the ligands, Bmp4, is under stringent control, subject to regulation by extracellular binding proteins that quench its activity. To date, three Bmp binding proteins have been identified in vertebrates. The first, follistatin, has been reported to bind Bmp7 *in vitro* (Yamashita et al., 1995), but its primary function *in vivo* is probably to inhibit activin. The second, noggin, is a relatively small, heparin-binding glycoprotein that binds Bmp2 and 4, as well as Bmp7, at lower affinity (Zimmerman et al., 1996). The third, chordin, is a large multidomain protein that binds Bmp2 and 4 in vitro (Francois and Bier, 1995; Piccolo et al., 1996). Whether other members of the Bmp family have affinity for these three factors, and whether other binding proteins exist, is under investigation.

In the *Xenopus* embryo, chordin is initially expressed around the dorsal lip and in the notochord, together with noggin and follistatin (Sasai et al., 1995). Chordin and noggin are thought to inhibit and limit the activity of Bmp4, which is expressed in a reciprocal domain throughout the animal hemisphere. Since Bmp4 supresses the differentiation of ectoderm into neuroepithelium, and promotes the differentiation of the marginal zone cells into ventral mesoderm, its antagonists have the opposite activity *in vivo*, acting as dorsalizing factors promoting the formation of neural tissue and dorsal mesoderm (see Graff, 1997, for review). The discovery that chordin is related to the product of the *Drosophila* gene, short gastrulation or *sog*, is particularly exciting because there is strong genetic evidence that *in vivo* sog antagonizes the activity of Decapentaplegic (Dpp), the insect homolog of Bmp2/4. This suggests that the antagonistic relationship between Dpp and sog, and between Bmp4 and chordin, has been evolutionarily conserved (Holley et al., 1995). Moreover, such "agonist-antagonist" diads may represent a general mechanism of regulating the function of many classes of

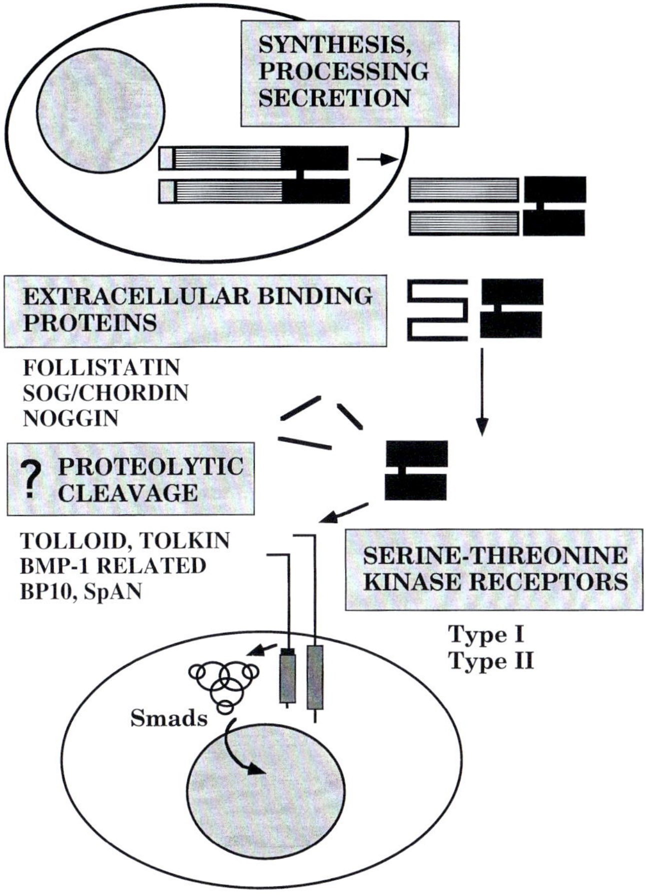

Fig. 1. Schematic representation of different processes involved in intercellular signaling by Bmp family members. Bmp precursor proteins undergo extensive post translational modification to yield an extracellular, C-terminal dimer. This may associate with specific binding proteins that quench its biological activity, and either sequester the ligand in extracellular compartments or form a reservoir of potentially active protein. The binding proteins may be cleaved by specific proteases, releasing the ligand and making it available for interaction with transmembrane serine-threonine kinase receptors. Downstream signaling from the receptors involves multimeric smad proteins that are phosphorylated and move from the cytoplasm to the nucleus. At the DNA level, smads may form part of a complex with DNA binding proteins, and thereby directly or indirectly activate or repress gene transcription. For details and references, see text.

signaling molecules *in vivo*. The possibility that further layers of complexity accompany extracellular Bmp signaling comes from studies in *Drosophila* of the genes *tolloid* and *tolkin*, which counteract the activity of *sog* and enhance the activity of *Dpp*. *Tolloid* and *tolkin* encode astacin proteases, and an attractive hypothesis is that *tolloid* normally cleaves sog protein in a controlled way, to release active Dpp (Childs and O'Connor, 1994; Finelli et al., 1995; Shimell et al., 1991). There is a large family of *tolloid*-related astacin proteases in both vertebrates and invertebrates [for example, sea urchin BP10 and SpAN (Hwang et al., 1994)]. One of these, Bmp1, functions as a collagen C-proteinase *in vitro* (Kessler et al., 1996), and also has a role in collagen processing during embryonic development (Suzuki et al., 1996). However, it is possible that in vivo other astacin proteases cleave proteins that bind to and inactivate Tgfβ family members.

Bmp proteins interact with target cells through transmembrane serine-threonine kinase receptors (see Massague et al., 1994, for review). These belong to two classes, Type I and Type II, that function as heteromers, possibly as tetramers. Currently, at least three Type I receptors have been identified that interact with known Bmps (Bmpr-IA, also known as Alk3 or Brk1, Bmpr-IB, also known as Alk6 or Brk2, and ActR-I or Alk2). *In vivo*, Alk3 and Alk6 receptor genes have different but overlapping expression domains (Dewulf et al., 1995). Three Type II receptors that bind to Bmps (Bmpr-II or Brk3, and ActR-II and ActR-IIB) have been identified. Both Type I and Type II subunits are required for downstream signaling, and it appears that the Type II kinase phosphorylates the Type I receptor to activate signal transduction. The affinities of different combinations of Type I and Type II receptors for homodimeric Bmp ligands vary in a complex way. Until more ligands have been purified and tested, and other potential receptors identified, the biological significance of these variable affinities is unclear. However, it is possible that the Type I receptors activate different responses in the same cell.

There have been rapid advances in our understanding of the downstream signaling cascade from Bmp receptors, catalyzed mainly by genetic studies in *Drosophila* and *C. elegans*. The current model is that activation of the Type I receptor kinase leads to the phosphorylation of proteins of the Smad family (Wrana and Attisano, 1996). For example, addition of Bmp2 to responsive cells leads to the phosphorylation of Smad1 and its movement from the cytoplasm to the nucleus (Hoodless et al., 1996). Smads, (Derynck et al., 1996) which are related to the products of both *Drosophila* mothers-against-dpp (*mad*) and *medea* genes, and *C. elegans sma* genes, are thought to form heteromeric complexes in the cytoplasm. One hypothesis is that different Type I receptor combinations activate different combinations of Smad proteins. In the nucleus, activated Smad proteins may either act as transcriptional activators (or repressors) or form part of a transcriptional complex with specific DNA binding factors.

III. Genetic Analysis of Bmp Function in Mice and Humans

One of the most powerful way of studying mammalian gene function in vivo is to analyze the phenotype of mutants. These may be generated spontaneously, by radiation, chemical, or insertional mutagenesis, or by homologous recombination in embryonic stem cells. Known mutations in the *Bmp* gene superfamily are summarized in Table I. To date, two spontaneous mouse mutants involving *Bmp* genes have been described: short ear ($Bmp5^{se}$) and brachypodism ($Gdf5^{bp}$). Short ear mice have mutations in the *Bmp5* gene; multiple recessive alleles have been identified over the years in mutant screens, and homozygotes show skeletal defects including loss or reduction in size of several small bones and cartilages, including the cartilage of the ear (King et al., 1994; Kingsley et al., 1992). These defects can be traced back to the formation of precartilagenous mesenchymal rudiments in which the *Bmp5* gene is expressed, although this gene is also transcribed in other

TABLE I.
Phenotypes of Mice Homozygous for Null Mutations in Bmp-Related Genes

Gene	Phenotype and major tissues affected	References
nodal	Embryonic lethal, mesoderm formation	Zhou et al., 1993; Conlon et al., 1994
Bmp2	Embryonic lethal, amnion, heart	Zhang et al., 1996
Bmp4	Embryonic lethal, mesoderm formation	Winnier et al., 1995
Bmp5	Viable, skeletal elements, kidney, lung	Kingsley et al., 1992
Bmp7	Perinatal lethal, kidney, eye, limb	Dudley et al., 1995; Luo et al., 1995
Bmp8b	Viable, males infertile	Zhao et al., 1996
Gdf5	Viable, joint abnormalities	Storm et al., 1994
Gdf8	Viable, muscle hypertrophy	McPherron et al., 1997
Gdf9	Viable, females infertile	Dong et al., 1996
Bmpr-1a	Embryonic lethal, mesoderm formation	Mishina et al., 1995

developing endochondral bones that do not show a defect. Homozygotes also show abnormalities in soft tissues where *Bmp5* is expressed; these include the lung, and ureters. Painstaking studies by Margaret Green showed that the phenotype of the *Bmp5*^{se} mutation varies, depending on the genetic background on which it is placed. In addition, heterozygotes also display a phenotype with low penetrance and expressivity, suggesting that the amount of protein made by cells is important for normal development. Brachypodism involves a recessive mutation in the *Gdf5* gene (Storm et al., 1994). This lead to the exciting finding that humans suffering from the conditions known as acromesomelic chondrodysplasia (OMIM #601146, #201250 and #200700) also have recessive or dominant mutations in the *GDF5* gene (Thomas et al., 1996). The mouse and human *Gdf5/GDF5* genes seem to be particularly important for the growth and differentiation of the different cell types in the joints of the limbs (Storm and Kingsley, 1996).

Over the past five years several *Bmp* family members have been inactivated by homologous recombination in ES cells; some homozygotes have no obvious phenotype, some show specific defects in adults, while others die perinatally or during embryonic development. For example, mice homozygous for a null mutation in *Bmp7* die perinatally, due to kidney failure. They also show defects in eye and limb development (Dudley et al., 1995; Luo et al., 1995). Embryos lacking a functional *Bmp2* gene die much earlier, around E9.5. In many of these homozygotes, the amnion fails to grow properly, and there are also major abnormalities in heart development (Zhang and Bradley, 1996). Most embryos homozygous for the closely related gene, *Bmp4*, die earlier in development, around the time of gastrulation; they fail to form embryonic mesoderm and do not develop beyond the egg cylinder stage (Winnier et al., 1995). About 20% of the *Bmp4* homozygotes do progress further, however, possibly due to compensation by the *Bmp2* gene, which is expressed in some of the same tissues as *Bmp4*, including the extraembryonic mesoderm of the yolk sac and amnion. This hypothesis is supported by the finding that none of the embryos homozygous for null mutations in the *Bmprla* gene, which encodes a Type I receptor that binds both Bmp2 and 4 (and 7) *in vitro,* develop beyond the egg cylinder stage *in vivo* (Mishina et al., 1995).

The variability of the *Bmp4* homozygous phenotype raised the possibility that genetic background could affect the penetrance and/or expressivity of the heterozygous phenotype, as in the case of *Bmp5*. This is supported by the finding that when the *Bmp4* null mutation is placed on the C57BL/6 inbred background, but not on 129/Sv, abnormalities are seen in heterozygotes, at a low but consistent frequency (Dunn et al., 1997). Abnormalities include unilateral cystic kidney, craniofacial defects, circling behavior, and preaxial polydactyly of the right hind limb. It is unlikely that these abnormalities are due to spontaneous mutations in other genes closely linked to *Bmp4* since they have been seen on mouse lines derived from two independent sources of ES cells, and involve tissues in which *Bmp4* is expressed. A more likely explanation is that C57BL/6 alleles for proteins involved in the synthesis, secretion, processing, availability, and downstream signaling of Bmp4 differ in efficiency from those in other strains of mice, such as 129/Sv. These observations may, in the future, be helpful not only for identifying key components of the Bmp signaling system but also for understanding the variable phenotype of members of human families carrying mutations in *Bmp* genes.

The fact that mice heterozygous for null mutations in *Bmp4* show abnormalities suggests that the amount of ligand produced in certain tissues is important for normal development. In the future it should be possible to test this hypothesis by making hypomorphic alleles of *Bmp4*. Compound heterozygotes for both the null and hypomorphic alleles should then show exacerbated phenotypes. In addition, if *Bmp2* normally compensates for the absence of *Bmp4* in some tissues, then double heterozygotes carrying null mutations in both genes should have more severe abnormalities than either heterozygote alone. Preliminary studies on *Bmp2;Bmp4* double heterozygotes suggest that this is indeed the case. Most double heterozygotes die before E16, but the cause of death is not yet known (Yuji Mishina and Richard Behringer, personal communication).

IV. Role of BMPs in Dorsal Forebrain Development

Analysis of mutant phenotypes can provide important information about the role of *Bmp* genes in vertebrate embryonic development. However, the role of a gene may be difficult to uncover in all tissues using this approach, due to compensation by related genes co-expressed in the same or nearby cells. Other experimental approaches are therefore necessary. These include injection of RNA into *Xenopus* oocytes, misexpression using heterologous promoters in transgenic mice and retroviral driven expression in chick embryos, and the addition of purified protein to primary cultures or cell lines. These techniques have provided important information about the role of Bmps in neural development.

Bmp family members have been implicated in several different processes in neurogenesis, beginning with the establishment of the neural plate. For example, around the time of gastrulation, Bmp4 inhibits neural differentiation in the ectoderm of the *Xenopus* animal hemisphere, and its activity has to be counteracted by dorsalizing factors such as noggin and chordin (for references, see previous section). Later in neurogenesis, Bmp4 and Bmp7, expressed in the dorsal surface ectoderm immediately adjacent to the neural tube, have been implicated in the dorsal-ventral patterning of the spinal cord (Liem et al., 1995). Most of this latter work has been done with the chick embryo, where the addition of the purified Bmps to undifferentiated neurectoderm from the intermediate region of the developing spinal cord is sufficient to induce genes normally expressed in dorsal cells. Recently, studies in the mouse embryo have suggested that Bmps also play a crucial role in the D-V patterning of the forebrain (Furuta et al., 1997). *In situ* hybridization studies have shown that by E10.5 there are high levels of expression of at least four *Bmp* genes (*Bmp4, 5, 6,* and *7*) in the neurectoderm in the dorsal midline of the telencephalon and anterior diencephalon, as well as in the roof of the hindbrain. Although some *Bmp*s are also expressed in the roof of the midbrain and spinal cord, it is quite striking how the transcripts of multiple *Bmp*s are co-localized in a restricted region of the dorsal forebrain (Fig. 2). As the forebrain develops, the neuroectoderm of the lateral telencephalon both thickens and expands dramatically, giving rise to the cerebral hemispheres. By contrast, the dorsal neuroepithelium, where the *Bmp*s are co-expressed, remains thin, and eventually gives rise to, among other things, the choroid plexus. This tissue, which eventually extends as frond-like projections into the forebrain ventricles, consists of a secretory epithelium and a layer of highly vascularized mesodermal cells. Its function is to produce and secrete cerebrospinal fluid. *In situ* hybridization studies show that *Bmp4, 5, 6* and *7* are transcribed in the epithelium of the choroid plexus, raising the interesting possibility that the proteins are secreted into the cerebrospinal fluid.

The fact that the neuroepithelium of the dorsal midline of the forebrain remains thin, while the lateral walls expand and thicken, suggests that there are differential rates of cell proliferation and/or cell death between the two regions *in vivo*. This is indeed the case, as shown by measuring the percentage of cells that incorporate BrdU into DNA during a two-hour labeling period at E10.5. Cells in the dorsal midline show on average about 2/3 the level of cell proliferation of cells in the lateral regions (Fig. 3). Apoptosis was also studied using the TUNEL method, which specifically labels "frag-

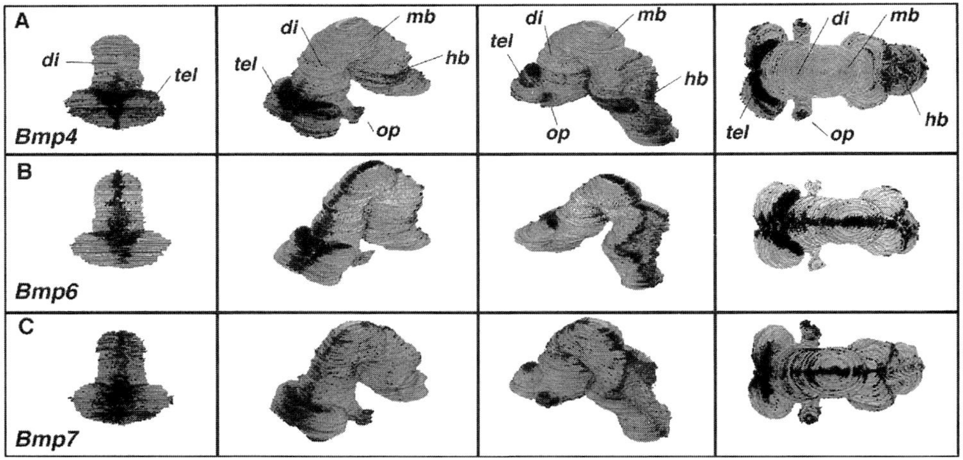

Fig. 2. Expression of Bmps in the brain of a 10.5 d.p.c. embryo. Three-dimensional reconstruction of the in situ hybridization images of *Bmp4* (row A), *Bmp6* (row B) and *Bmp7* (row C) viewed from the front, laterally from the front, laterally from the back and from the top (left to right). Note that only expression in the neuroepithelium is shown. Black represents the *in situ* signals of the *Bmps*. Abbreviations: di, diencephalon; mb, midbrain; hb, hindbrain; op, optic cup/stalk; tel, telencephalon.

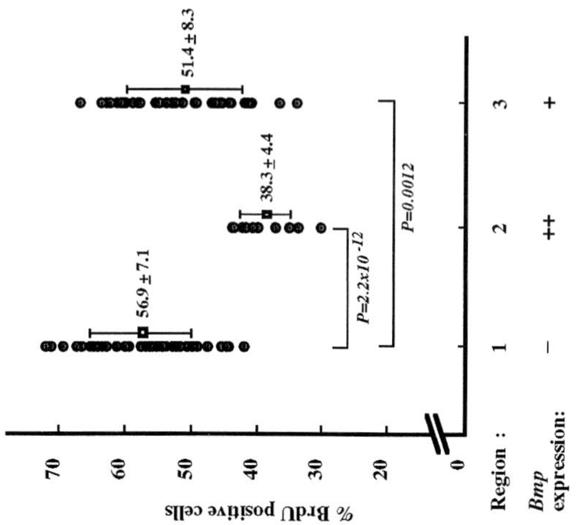
% BrdU positive cells
56.9 ± 7.1
38.3 ± 4.4
51.4 ± 8.3
P=2.2x10 -12
P=0.0012
Region :
Bmp expression:

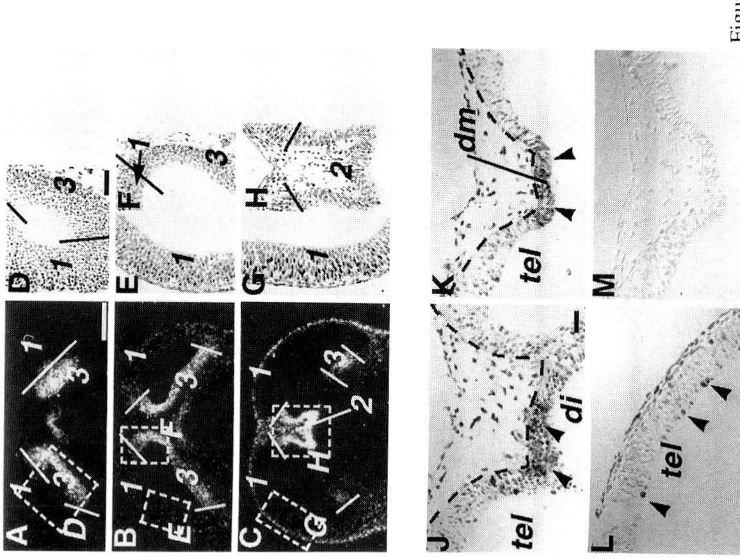
tel
di
dm
tel
tel

Figure 3.

mented" DNA in the process of degradation, i.e., a relatively late stage in the cascade of events associated with programmed cell death. In this case, there were many more positive cells in the dorsal midline of the anterior diencephalon and forebrain, compared with the lateral and ventral regions. These *in vivo* observations raised the possibility that reduced cell proliferation and increased apoptosis in the dorsal midline are causally related to the high level of expression of *Bmps*. Analysis of mutant embryos does not throw much light on this hypothesis; most *Bmp4* homozygotes die before neurogenesis, and no obvious defects have been reported in forebrain development in *Bmp5, 6* or *7* null mice, presumably due to functional compensation between genes. An explant culture was therefore established to test the effect of Bmps on forebrain neuroepithelium. Small pieces dissected at E10.5 from the lateral telencephalic vesicles, where *Bmps* are not expressed, were separated from the overlying mesenchyme and cultured on sterile filters. Beads soaked in either purified Bmp proteins or bovine serum albumin were applied to the tissue, and after various times the explants were fixed and assayed for specific gene expression by *in situ* hybridization, and for cell

---

Fig. 3. Cell proliferation and cell death in the telencephalon of 10.5 d.p.c. embryos. **A–C:** Expression of *Bmp6* (A), *Bmp5* (B), and *Bmp4* (C) in the telencephalon of a 10.5 d.p.c. embryo (dorsal to the top). The dorsal telencephalon is divided, according to the level of *Bmp* expression; (1) the lateral region, where *Bmps* are not expressed, (2) the dorsal midline, and (3) the dorsomedial region, where *Bmps* are expressed. **D–H:** Examples of BrdU antibody staining in the regions shown in A–C. **I:** Percent BrdU positive cells in the neuroectoderm was determined by the number of positive cells/total cells in each region. The data are from a total of 41,400 cells counted on six different sections from each of three embryos. Each dot represents the value from one photomicrograph containing 200–500 cells. Mean values and standard deviations from each group are also shown. The differences in cell proliferation in *Bmp* non-expressing and expressing regions are statistically significant by Student's *t*-test ($P<0.01$). **J,K:** Programmed cell death detected by TUNEL (arrowheads) in the dorsal midline of the forebrain region, at the diencephalon/telencephalon junction (J), and at the dorsal roof between telencephalic hemispheres (K). The border between the head mesenchyme and neuroectoderm is indicated by hatched lines. **L:** The lateral telencephalon in the same section as the one shown in (K). TUNEL-positive cells are barely detectable in the lateral and ventral telencephalon at this stage (arrowheads). **M:** A section near the one shown in (K) incubated with heat-inactivated terminal deoxynucleotidyl transferase as a negative control. Abbreviations: di, diencephalon; dm, dorsal midline of the telencephalon; tel, telencephalon. Bar=200 μm (A–C), 50 μm (D–H), 25 μm (J–M).

proliferation and apoptosis. These studies showed that beads soaked in Bmp4 (or Bmp2) induce a zone of decreased cell proliferation around them, as measured by reduced incorporation of BrdU after 18 and 24 hours culture (Fig. 4G). The same zone also shows increased TUNEL labeling by 18 hours (Fig. 4H). This finding supports the hypothesis that Bmp4 negatively regulates cell proliferation in the dorsal forebrain neuroepithelium *in vivo* and induces apoptosis in at least a proportion of the cells. In several other embryonic systems (for example, the tooth bud, spinal cord, facial primordium and limb), Bmp4 signaling has been associated with an increase in the expression of the homeobox genes *Msx1* and *Msx2* in target cells (Chen et al., 1996; Kratochwil et al., 1996; Liem et al., 1995; Vainio et al., 1993). The normal expression of *Msx1* was therefore studied in the developing forebrain, and high levels of RNA were found in the dorsal midline, coincident with the region of *Bmp* co-expression, but not in the lateral telencephalon. However, using the explant culture system, Bmp4-soaked beads were able to induce *Msx1* expression in the lateral neuroepithelium, in a zone of about 10–15 cell diameters around the bead (Fig. 4 A–D). Time course experiments show that *Msx1* expression precedes the decreased BrdU incorporation and increased TUNEL labeling in the cells around the bead. Whether *Msx1* directly regulates processes involved in cell cycle progression and/or apoptosis is under investigation.

These results support the hypothesis that Bmp4 plays a role in the development of the dorsal forebrain. However, many important questions remain unanswered. Some of these are specific to the dorsal forebrain, but others illustrate more general problems that have to be addressed in all embryonic tissues where Bmps are implicated in controlling growth and patterning. For example, there does not appear to be a gradient of *Msx1* expression around the beads, suggesting that the dose of Bmp4 has to exceed a threshold for inducing *Msx1*. Are other genes induced in the neurectoderm cells at lower threshold levels of Bmp4, and does the protein form a morphogenetic gradient *in vivo*? Are Bmp4-binding proteins such as chordin and noggin secreted by cells in the dorsal midline, and do they regulate the availability and/or diffusion of active Bmp4 and/or protect a proportion of the cells from the growth inhibitory effects of high levels of Bmps, so that they can go on to differentiate into secretory epithelial cells of the choroid plexus? Likewise, *Bmp4* RNA is normally expressed at high levels in the mesenchyme surrounding the lateral forebrain, but the cells do not appear to be exposed to active protein. Do the extracellular components of the basal lamina nor-

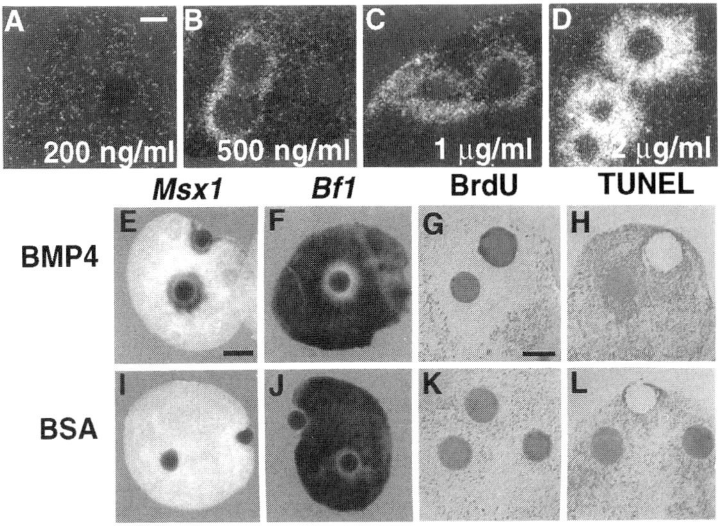

Fig. 4. Effect of BMP4 protein in explant cultures. **A–D:** *In situ* hybridization on telencephalic explants showing dose dependency of the induction of *Msx1* expression by BMP4 protein after 24 hours of culture. **E,F:** Whole mount *in situ* hybridization of the explants showing induction of *Msx1* expression (E) and suppression of *Bf1* expression (F) at 24 hours. **G:** Suppression of cell proliferation by BMP4 in the explants at 24 hours. **H:** In the adjacent section of (G), significant accumulation of TUNEL-positive cells is notable around the BMP4-carrying beads. **I:** BSA/PBS does not induce *Msx1* expression even after 48 hours of culture. **J:** BSA/PBS has no effect for *Bf1* expression under the same conditions. **K:** BSA/PBS beads do not affect cell proliferation. **L:** A section near the one shown in (F), exhibiting random distribution of TUNEL-positive cells around the BSA/PBS beads. Bar=50 μm (A–D), (E,F,I,J), (G,H,K,L).

mally bind Bmps and prevent their access to the neuroepithelium. What is the significance of the fact that multiple *Bmp* genes are co-expressed in the dorsal midline? One possibility is that members such as Bmp4 and Bmp7 form heterodimers that are more potent or have different activities than either homodimer. Certainly, it appears that Bmps 5, 6 and 7 are not obligatory for forebrain development *in vivo* since none of the mice homozygous for null mutations in these genes have significant defects in this tissue. The possibility of functional compensation between genes is currently being tested by making double mutants. Finally, there is the interesting question of the specificity of the effect of Bmp4 on different regions of the neuroepithelium. Bmp4 can induce apoptosis in the neuroepithelium of the hindbrain, specifically in rhombomeres 3 and 5 (Graham et al, 1997). However, Bmp4 does not induce apoptosis in rhombomere 4. The molecular basis for this differential sensitivity is not yet known.

V. Future Directions

Mutant analysis, and functional assays involving techniques such as embryonic tissue recombinations, *in vitro* culture, and ectopic application of purified protein have already provided us with a wealth of information about the role of Bmps in vertebrate development. This applies not only to the function of the ligands, but to the whole system of binding proteins, activators, receptors, and components of the downstream signaling cascade. Nevertheless, there is still much that needs to be learned. In the future, more sophisticated genetic approaches, such as tissue-specific gene inactivation, introduction of specific amino acid alterations into proteins, and identification of modifier genes in mutant supressor and enhancer screens, will help to fill the gaps.

Learning everything about how Bmps function is obviously important, but it only provides insight into one aspect of an extremely complex puzzle. Perhaps the most exciting direction for the future is to understand how Bmps work together with members of other signaling protein families, such as the Wnts, Fgfs, and Hhs, to co-ordinate embryonic growth and patterning. Are there general rules, that have been conserved during evolution? Are there mechanisms by which pathways from different signaling proteins converge in the same cell, to enhance or counteract one another at the level of gene transcription? Cells in embryos are often exposed to more than one growth factor or signaling molecule at a time, and behave differently if they receive input simultaneously from two or more classes of receptors, compared with only one. At some level the incoming information from signaling systems has to be integrated, and one place where this may happen is at the regulatory regions of target genes. For example, it has recently been shown that an upstream region of the *Ubx* gene in *Drosophila* contains both Dpp and Wingless response elements in close proximity (Riese et al. 1997). Similarly, *cis*-acting response elements for both Wnt and activin/BVg have been identified in the Xenopus *goosecoid* gene (Watabe et al., 1995; Schwyter et al., 1995). Future studies will provide insight into how transcription factors binding to these elements co-ordinate the trancriptional response of cells to multiple positional signals, including Bmps.

ACKNOWLEDGMENTS

I gratefully acknowledge the enthusiastic contribution of my laboratory colleagues to many of the experiments and ideas discussed here. In particular, Drs. Karen Lyons and C. Michael Jones contributed to the early work on Bmps, and Manfred Blessing, Glenn Winnier, Nancy Wall, Patricia Labosky, Noboru Suzuki, Guang Zhao, Yasuhide Furuta, N. Ray Dunn, and Saverio Bellusci have all subsequently carried the flag. Finally, Dr. Christopher Wright has provided hours of stimulating discussion and advice over many years. Work on Bmps in my lab was funded by grants from the National Institutes of Health, and my laboratory has also been supported by the Howard Hughes Medical Institute.

REFERENCES

Chen, Y., Bei, M., Woo, I., Satokata, I., and Maas, R. (1996). *Development* **122,** 3035–3044.
Childs, S.R., and O'Connor, M.B. (1994). *Dev. Biol.* **162,** 209–220.
Conlon, F.L., Lyons, K.M., Takaesu, N., Barth, K.S., Kispert, A., Herrmann, B., and Robertson, E.J. (1994). *Development* **120,** 1919–1928.
DeRobertis, E.M. (1997). *Nature* **387,** 25–26.
Derynck, R., Gelbart, W.M., Harland, R.M., Heldin, C.-H., Kern, S.E., Massague, J., Melton, D.A., Mlodzik, M., Padgett, R.W., Roberts, A.B., Smith, J., Thomsen, G.H., Vogelstein, B. and Wang, X.-F. (1996). *Cell* **87,** 173.
Dewulf, N., Verschueren, K., Lonnoy, O., Moren, A., Grimsby, S., Vande Spiegle, K., Miyazono, K., Huylebroeck, D. and ten Dijke, P. (1995). *Endocrinology* **136,** 2652–2663.
Dong, J., Albertini, D.F., Nishimori, K., Kumar, T.R., Lu, N., and Matzuk, M.M. (1996). *Nature* **383,** 531–534.
Dudley, A.T., Lyons, K.M., and Robertson, E.J. (1995). *Genes Dev.* **9,** 2795–2807.
Dunn, N.R., Winnier, G.E., Hargett, L.K., Schrick, J.J., Fogo, A.B., and Hogan, B.L.M. (1997). *Dev. Biol.* **188,** 235–247.
Finelli, A.L., Xie, T., Bossie, C.A., Blackman, R.K., and Padgett, R.W. (1995). *Genetics* **141,** 271–281.
Francois, V., and Bier, E. (1995). *Cell* **80,** 19–20.
Furuta, Y., Piston, D.W., and Hogan, B.L.M. (1997). *Development* **124,** 2203–2212.
Graff, J. (1997). *Cell* **89,** 171–174.
Graham, A., Francis-West, P., Brikell, P., and Lumsden, A. (1994). *Nature* **372,** 684–686.
Griffith, D.L., Keck, P.C., Sampath, T.K., Rueger, D.C., and Carlson, W.D. (1996). *Proc. Natl. Acad. Sci. U.S.A.* **93,** 878–883.
Harland, R.M. (1994). *Proc. Natl. Acad. Sci. U.S.A.* **91,** 10243–10246.
Hogan, B.L.M. (1996). *Genes Dev.* **10,** 1580–1594.
Hoodless, P.A., Haerry, T., Abdollah, S., Stapleton, M., O'Connor, M.B., Attisano, L., and Wrana, J.L. (1996). *Cell* **85,** 489–500.
Holley, S.A., Jackson, P.D., Sasai, Y., Lu, B., De Robertis, E.M., Hoffmann, F.M., and Ferguson, E.L. (1995). *Nature* **376,** 249–253.
Hwang, S.-P. L., Partin, J.S., and Lennarz, W.J. (1994). *Development* **120,** 559–568.
Kessler, E., Takahara, K., Biniaminov, L., Brusel, M., and Greenspan, D.S. (1996). *Science* **271,** 360–362.

King, J., Marker, P.C., Seung, K.S., and Kingsley, D.M. (1994). *Dev. Biol.* **166,** 112–122.

Kingsley, D.M. (1994). *Genes Dev.* **8,** 133–146.

Kingsley, D.M., Bland, A.E., Grubber, J.M., Marker, P.C., Russell, L.B., Copeland, N.G., and Jenkins, N.A. (1992). *Cell* **71,** 399–410.

Kratochwil, K., Dull, M., Farinas, I., Galceran, J., and Grosschedl, R. (1996). *Genes Dev.* **10,** 1382–1394.

Liem, J.K.F., Tremml, G., Roelink, H., and Jessell, T.M. (1995). *Cell* **82,** 969–979.

Luo, G., Hofmann, C., Bronckers, A.L.J.J., Sohocki, M., Bradley, A., and Karsenty, G. (1995). *Genes Dev.* **9,** 2808–2820.

Massague, J., Attisano, L., and Wrana, J.L. (1994). *Trends Cell Biol.* **4,** 172–178.

McPherron, A.C.. and Lee, S.-J. (1996). {{The Transforming Growth Factor β Superfamily." JAI Press Inc., Greenwich, Conn.

McPherron, AC., Lawler, AM., and Lee, S-J. (1997). *Nature* **387,** 83–90.

Mishina, Y., Susuki, A., Ueno, N., and Behringer, R. R. (1995). *Genes Dev.* **9,** 3027–3037.

Piccolo, S., Sasai, Y., Lu, B., and De Robertis, E. M. (1996). *Cell* **86,** 589–598.

Riese, J., Yu, X., Munnerlyn, A., Eresh, S., Hsu, S-C., Grosschedl, R., and Bienz, M. (1997). *Cell* **88,** 777–788.

Sasai, Y., Lu, B., Steinbeisser, H.. and De Robertis, E. M. (1995). *Nature* **376,** 333–337.

Schwyter, D.H., Huang, J-D., Dubnicoff, T., and Cowey, A.J. (1995). *Mol. Cell. Biol.* **15,** 3960–3968.

Shimell, M.J., Ferguson, E.L., Childs, S.R., and O'Connor, M.B. (1991). *Cell* **67,** 469–481.

Storm, E.E., Huynh, T.V., Copeland, N.G., Jenkins, N.A., Kingsley, D.M., and Lee, S.-J. (1994). *Nature* **368,** 639–643.

Storm, E.E., and Kingsley, D.M. (1996). *Development* **122,** 3969–3979.

Suzuki, N., Labosky, P., Furuta, Y., Hargett, L., Dunn, R., Fogo, A., Takahara, K., Peters, D., Greenspan, D., and Hogan, B. (1996). *Development* **122,** 3587–3595.

Thomas, J.T., Lin, K., Nandedkar, M., Camargo, M., Cervenka, J. and Luyten, F. (1996). *Nature Genet.* **12,** 315–317.

Vainio, S., Karavanova, I., Jowett, A., and Thesleff, I. (1993). *Cell* **75,** 45–58.

Watabe, T., Kim, S., Candia, A., Rothbacher, U., Hashimoto, C., Inoue, K. and Cho, K. W. Y. (1995). *Genes Dev.* **9,** 3038–3050.

Winnier, G., Blessing, M., Labosky, P. A., and Hogan, B. L. M. (1995). *Genes Dev.* **9,** 2105–2116.

Wrana, J.L., and Attisano, L. (1996) *Trends Genet.* **12,** 493–496.

Yamashita, H., ten Dijke, P., Huylebroeck, D., Sampath, T.K., Andries, M., Smith, J.C., Heldin, C.-H., and Miyazono (1995). *J. Biol. Chem.* **130,** 217–226.

Zhang, H., and Bradley, A. (1996). *Development* **122,** 2977–2986.

Zhao, G-Q., Deng, K., Labosky, P.A., Liaw, L., and Hogan, B.L.M. (1996). *Genes Dev.* **10,** 1657–1669.

Zhou, X., Sasaki, H., Lowe, L., Hogan, B.L.M., and Kuehn, M.R. (1993). *Nature* **361,** 543–547.

Zimmerman, L.B., De Jesus-Escobar, J.M., and Harland, R.M. (1996). *Cell* **86,** 599–606.

THE MOLECULAR BASIS OF IRON METABOLISM

RICHARD D. KLAUSNER AND TRACEY ROUAULT

National Cancer Institute and National Institute of Child Health and Human Development, Bethesda, Maryland

I. INTRODUCTION

The maintenance of the non-equilibrium state central to life is made pos sible by successful capture and transduction of chemical free energy. Metals play a central role in many biological processes including those involved with energy transduction; one of these metals, whose flexible coordination chemistry and accessible redox reactivity have made it indispensable, is iron. Indeed, some have speculated that it was on the surface of iron pyrite—fool's gold—that the central catalysis of metabolism began, even before these metabolic pathways were sequestered within cellular compartments (reviewed in Maden, 1995; Wachtershauser, 1992). Virtually all forms of life require iron. From a ferrocentric point of view, life began in a ferrous paradise; ferrous iron was plentiful and readily available and our dependence on iron was acquired early. But, paradise was not to last. One of the first and perhaps most cataclysmic examples of the ability of life forms to impact upon the environment was the oxygenation of the atmosphere. While we are understandably quite enamored of oxygen, the consequences of the oxygenation of the atmosphere for cells that had become dependent on iron was profound. Our ancestors were consigned to ferric purgatory as the solubility of iron dropped from 0.1 M in the ferrous form at neutral pH to 10^{-17} M, concentrations incompatible with the needs of any known cell. The presence of oxygen not only reduced the concentration of ferrous iron, but also created tremendous toxic potential because iron can react with various oxygen species to generate highly reactive hydroxyl radicals (Gutteridge, 1994).

Iron is rate-limiting for growth in many ecological niches. Certain opportunistic infections are particularly virulent because they can extract otherwise unavailable iron from within an organism (reviewed in Weinberg, 1993).

The Harvey Lectures, Series 92, pages 99–112

The yield of agricultural crops is limited by the ability of the roots to extract iron from the rhizosphere (Eide and Guerinot, 1996). It has been claimed that the ability of the biosphere to detoxify CO_2 is limited by the biomass of zooplankton in the oceans and that biomass, in turn, is limited by the amount of bioavailable iron. For biological systems, iron is a two-edged sword: life forms must subsist within a narrow range of iron concentrations, avoiding iron deficiency on the one hand and iron toxicity in the other.

This paper covers 17 years of research on the central metabolic pathways of iron homeostasis. Not surprisingly, along the way, iron metabolism has provided several intriguing forays into areas of biology as diverse as intracellular traffic, RNA-based gene regulation, and environmental sensing.

II. The Transferrin Cycle and the Uptake of Iron

Our studies on iron metabolism began with the attempt to understand how mammalian cells acquire iron. Most mammalian cells acquire iron from the ferric chelator (solubilizer) transferrin, a serum protein that tightly binds two moles of iron per mole of protein. In the late 1970s, it was appreciated that all proliferating cells have transferrin receptors that bind ferric transferrin with high affinity. Using cells in culture, we demonstrated that the transferrin–iron complex was internalized via coated pit-mediated endocytosis, and we were able to show that this complex was delivered to an acidic compartment that was identified as the endosome. The acidification was essential for the release of iron from the transferrin into the lumen of the endosome. At the time, it was believed that endocytosis was a one-way path that delivered internalized ligand to lysosomes for degradation. We showed that the apo-transferrin generated in the acidic endosome remained bound to the transferrin receptor and that the receptor and the apo-protein recycled to the plasma membrane where the return to neutral pH resulted in the rapid release of the apo-transferrin to allow this "transferrin cycle" to begin again. At the time, we did not know how the iron released from transferrin was transported from the endosome into the cytosol, but we were able to demonstrate that this iron transport was efficient (Klausner et al., 1983).

Once in the cytosol, the iron was distributed to a variety of targets including ferritin, a protein composed of 24 subunits that co-assemble to form a hollow sphere into which iron is deposited. We demonstrated two things about this process: 1) in these cultured cells, the iron that entered ferritin remained sequestered for long periods of time and did not appear to be readily available for reutilization; and 2) the percentage of the iron delivered from transferrin that was sequestered in ferritin was directly proportional to the amount of ferritin in the cell (Mattia et al., 1986).

We next turned our attention to identifying the mechanisms that maintain iron homeostasis, in part because we thought that unique sensing and regulatory mechanisms might exist in a system in which a delicate balance must be maintained. We demonstrated that the rate of flux of iron into cells was directly proportional to the number of transferrin receptors. Furthermore, the fraction of the iron fluxed into the cells through the receptor that was available for use in general metabolism was inversely proportional to the amount of ferritin in the cells (Mattia et al., 1984). The three gene products involved, the transferrin receptor and the two ferritin genes (H and L), were highly and rapidly regulated in response to changes in iron availability. When cells were iron-deprived, synthesis of transferrin receptors increased, allowing cells to capture more external ferric transferrin. Simultaneously, ferritin synthesis decreased, assuring that the iron taken up in the transferrin cycle was maximally available for use in general metabolism and minimally sequestered.

III. Coordinate Regulation of Iron Uptake and Availability

We next turned our attention to trying to understand the molecular basis for the coordinate but opposite regulation of the two ferritin gene products and the transferrin receptor. Previous work by Hamish Munro and colleagues (reviewed in Munro and Eisenstein, 1989) had indirectly shown that ferritin was regulated at the translational level in response to changes in iron, such that ferritin biosynthesis increases in iron-replete cells and is repressed in cells depleted of iron. Once the cDNAs for ferritin were cloned, Northern blot analysis revealed that there was no change in cytosolic ferritin mRNA ferritin levels that could account for significant changes in protein synthesis rates (Rouault et al., 1987). In the case of the transferrin receptor, manipulations of intracellular iron levels produced changes in the rate of protein synthesis that were directly proportional to changes in cytosolic mRNA levels. In iron-replete cells, levels of TfR mRNA decreased, whereas in iron-depleted cells, levels of TfR mRNA increased, and these changes in mRNA levels were attributable to iron-dependent changes in the stability of the mRNA itself (reviewed in Casey et al., 1989).

We began our search for the components of the iron regulatory process by cloning the full-length human ferritin H chain gene; we identified a small region (approximately 30 nucleotides) near the 5′ end of the 212 nucleotide 5′ UTR that was necessary for iron-dependent control of ferritin translation (Hentze et al., 1987b). This small sequence was sufficient to confer iron-dependent translational control upon unrelated transcripts when transferred to the 5′ end of heterologous transcripts. The region of the RNA identified as important by deletional analysis was predicted to form a hairpin loop that we called an iron-responsive element or IRE (Hentze et al., 1987a). Several point mutations in this IRE eliminated iron-dependent translational control, and the full range of translational control observed in normal cells could be attributed to the function of this small *cis*-acting element (Caughman et al., 1988). IRE stem loop structures were identified in all known ferritins and their location in the 5′ UTR of each transcript was conserved.

The regulation of the transferrin receptor was more complex. We and others demonstrated that expression of the transferrin receptor was post-transcriptionally regulated, and we determined that the regulatory region was contained within a 680-nucleotide region in the middle of the 2.3 kb 3′ UTR found within the transferrin receptor transcript (Casey et al., 1989; Mullner et al., 1989). Both visual and computer-based folding of this regulatory region revealed a highly structured segment of RNA that contained five stem-loop structures; these stem loops were strikingly similar to the consensus IRE derived from comparison of all known ferritin regulatory IREs. Indeed, several of these transferrin receptor mRNA 3′ UTR potential IREs were cloned into the 5′ UTR of reporter genes and were shown to confer ferritin-like translational control upon the recipient transcripts (Casey et al., 1988). The complex TfR regulatory region could be deconvoluted into two interwoven regulatory elements—the IRE elements and a sequence element that facilitated rapid degradation of the transcript in iron-depleted cells. We subsequently demonstrated that this second component, an RNA instability element, resulted in a site-specific endonucleolytic cleavage that initiated the degradation of the TfR mRNA (Binder et al., 1994).

Thus, we were able to identify an evolutionarily conserved specific stem-loop structure that endowed three iron-regulated messages with the ability to sense iron levels, transducing that sensing into an altered fate of the RNA. In the case of 5′ UTR IREs that are found in ferritins, this element produced translational control. In the case of the transferrin receptor, the IREs were embedded in a 3′ UTR RNA instability element in which they were capable of modulating mRNA half-life.

IV. The Discovery of IRE-Binding Proteins

Several studies suggested that the presence of an IRE in the 5′ UTR of a transcript allowed the initiation of translation to be repressed under conditions of iron deprivation. At the time of these investigations, there were no known regulatory RNA-binding proteins that mediated alterations in the fate of specific RNAs. To better understand the regulation observed, we used the IRE to search for specific *trans*-acting factors. Munro's lab had identified an IRE-specific-binding protein in the cytosol of all cells using a gel mobility shift of labeled IRE RNA (Leibold and Munro, 1988). We used this assay to determine that the binding activity of the protein that bound to IREs was high affinity (10 to 30 picomolar Kd) and specific for the RNA stem loop that functioned as a translational control element. Perhaps most revealing was the observation that the level of the binding activity found in the cytosol, as assessed in cell lysates, reflected the iron status of the cell (Rouault et al., 1988). When cells were iron-replete, there was little binding activity, whereas iron-starved cells had high levels of binding activity. The translation of ferritin that contained a functional IRE was inhibited in iron-deprived cells, and high IRE binding activity could be detected in the cytosol of these cells. We proposed a model in which IRE binding proteins function as simple repressors in the coordinate control of the cytosolic fate of iron-regulated transcripts (Fig. 1). According to this model, an IRE-binding protein was activated for RNA-binding under conditions of iron deprivation. When activated, it bound to any transcript containing a cognate binding site and repressed the function associated with the local region of the transcript. When an IRE was present within approximately 50 nucleotides of the cap site, the binding of IRE binding proteins blocked translation; it was subsequently demonstrated by Hentze and colleagues that the effect of binding of IRE binding proteins to the 5′ IRE was to prevent the binding of translational initiation factors to the cap site (Gray and Hentze, 1994). When IRE binding proteins occupied the IREs interspersed within the 3′ UTR TfR instability element, degradation of the transcript was inhibited, and TfR mRNA accumulated, resulting in an increase of TfR biosynthesis. In this way, the regulation of the RNA-binding activity of a specific cytosolic protein in response to altered iron levels could simultaneously and coordinately regulate levels of proteins involved in the uptake or sequestration of iron.

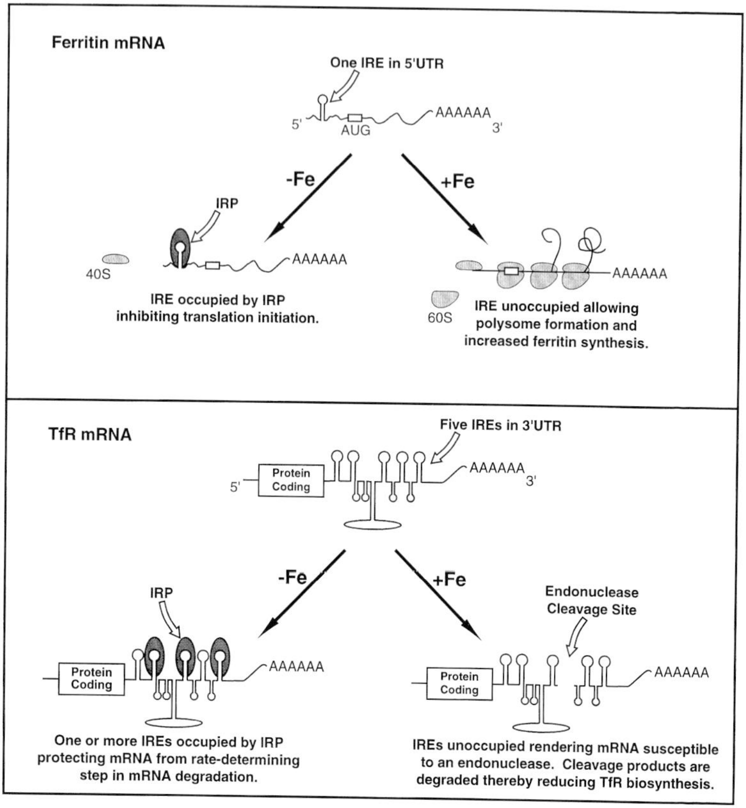

Fig. 1. Coordinate regulation of the expression of ferritin and the transferrin receptor. The 5′ UTR of ferritin mRNA contains a single iron-responsive element (IRE), whereas the 3′ UTR of the transferrin receptor contains five IREs. Changes in translation or transcript stability are depicted and are described in greater detail within the figure and in the text.

V. Cloning of IRE-Binding Proteins and Insights Into Mechanisms of Iron Sensing

To gain insight into the mechanism of iron-sensing of IRE-binding proteins, we purified and cloned the IRE-binding proteins. Two highly related proteins, now referred to as iron regulatory proteins 1 and 2 (IRP1, IRP2) were identified (Rouault et al., 1989, 1990). Database comparisons revealed significant amounts of information about these two proteins because they were both highly related to mitochondrial aconitase, an iron–sulfur protein that interconverts citrate and isocitrate in the citric acid cycle (Rouault et al., 1991, 1992). Although the sequence identity of IRP1 and mitochondrial aconitase was less than 30%, both are now known to be functional aconitases, and IRP1 is recognized as the sole source of aconitase activity in the cytosol (Kennedy et al., 1992).

When signals from the environment are sensed, it is expected that there will be a step in which there is a direct physical interaction between the substance that must be sensed and the sensor. In the case of IRP1, it was readily established that, like mitochondrial aconitase, a cubane cluster composed of iron and sulfur was ligated to the protein backbone through the interactions of irons in the cluster with cysteine sulfhydryl groups (Haile et al., 1992a,b; Kennedy et al., 1992). The demonstration that IRP1 directly binds iron led us to hypothesize that the bound iron, in the form of an iron–sulfur cluster, was the key to the regulatory function of IRP1. Many different studies have shown that IRP1 is a stable bifunctional protein; in iron-replete cells, IRP1 is an active enzyme, whereas in cells that are depleted of iron, IRP1 binds IREs with high affinity and lacks aconitase activity (reviewed in Klausner et al., 1993). We demonstrated that the form of IRP1 that binds IREs is apoprotein that is devoid of bound iron, whereas the enzymatically active form of IRP1 ligates an intact iron–sulfur cluster composed of four iron and four inorganic sulfur atoms.

The strong correlation between the status of the iron–sulfur cluster and function of the protein led us to perform a series of experiments designed to test whether alterations in the status of the iron–sulfur cluster were sufficient to change the function of IRP1. We reasoned that if the iron–sulfur cluster of IRP1 is the key to the switch from function as a binding protein to function as an active enzyme, then elimination of the ability to ligate an iron–sulfur cluster would turn the protein into a constitutive RNA binding protein, incapable of turning off RNA binding activity in iron-replete cells. Indeed, when the cysteines involved in ligation of the iron–sulfur cluster were mutagenized to serines, the protein functioned only as an IRE binding protein, regardless of the iron status of the cell (Philpott et al., 1993). To learn more about the RNA binding site, we performed cross-linking studies that showed that the IRE binds close to residues in the enzymatic active site (Basilion et al., 1994), and these observations were confirmed in other cross-linking studies (Swenson and Walden, 1994; Neupert et al., 1995). However, the theoretical predicted size of the IRE was such that it could not be accommodated within a cleft with dimensions similar to those measured in the crystal structure of mitochondrial aconitase. For that reason, we proposed a model in which the fourth domain of IRP1, which is connected to the first three domains of the protein by a potentially flexible hinge-linker, changes its position relative to the main body of the protein, opening the cleft and accommodating the RNA stem loop (Klausner and Rouault, 1993). Site-directed mutagenesis studies have shown that several arginines of the enzymatic active site are indispensable for high-affinity RNA binding, thus explaining why the enzymatic activity and RNA binding activities are mutually exclusive properties of the bifunctional IRP1 (Philpott et al., 1994).

The long half-life of IRP1 poses an interesting dilemma in understanding iron sensing, since binding activity of IRP1 changes within three hours when extracellular iron concentrations are altered, whereas the protein half-life exceeds 18 hrs (Tang et al., 1992). Much evidence now supports the conclusion that iron–sulfur clusters are relatively unstable prosthetic groups that spontaneously disintegrate upon exposure to oxidants (Beinert and Kiley, 1996; Rouault and Klausner, 1996). In mammalian cells, oxygen is used as the terminal electron acceptor in respiration, and highly reactive oxygen species are generated by the interaction of iron with oxygen species within cells. Thus, it is reasonable to propose that the iron–sulfur cluster of IRP1 is rapidly turned over and that a steady state is achieved in which the proportion of IRP1 that contains an iron–sulfur cluster reflects the concentrations of both cluster-destabilizing factors, as well as the concentrations of iron and sulfur that would be required to resynthesize the cluster. The majority of IRP1 would shift to the IRE binding form when sufficient iron for cluster reassembly is no longer available in the cytosol. Rapid turnover of the iron–sulfur prosthetic group of IRP1 would therefore be able to register changes in iron levels within hours, even in the absence of rapid turnover of the protein itself.

A second IRE binding protein (IRP2), cloned at the same time as IRP1, is also present in all cells that have been studied to date. The ability of IRP2 to bind IREs is high in iron-depleted cells and low in iron-replete cells, similar to IRP1. However, IRP2, unlike IRP1, is physically absent in iron-replete cells. Northern blot studies have revealed that mRNA levels of IRP1 do not change when iron levels are manipulated in cells. We performed pulse chase studies in order to determine whether changes in the rate of synthesis of degradation of IRP2 accounted for its absence in iron-replete cells. These studies revealed that the rate of synthesis of IRP2 was unaffected by iron status but that the rate of IRP2 degradation was increased markedly in iron-replete cells (Samaniego, et al., 1994; Guo et al., 1995). It was somewhat unexpected that a fundamental feature such as stability in the presence of iron would differ between two proteins that were otherwise strikingly similar, with sequence identity of 58%. However, there are several noteworthy differences between the two proteins that could be appreciated through studies of the sequence alignments. One difference was that IRP2 contained several mutations of residues that were known to be essential to the aconitase activity of mitochondrial aconitase. Indeed, we and others were able to demonstrate that IRP2, unlike IRP1, is not a functional aconitase (Samaniego, et al., 1994; Guo et al., 1994). A second striking difference was that IRP2 contained an insert of 73 amino acids relative to IRP1 in the first domain of the protein. Characterization of the genomic structure IRP2 revealed that this insert is derived from insertion of an extra exon into IRP2. We reasoned that this extra sequence of amino acids might be responsible for the acquisition of a new phenotype of IRP2—the ability to undergo iron-dependent degradation. To evaluate the role of the insert, we deleted the extra amino acids, and the deletion mutant, which retained the ability to bind IREs with high affinity, was now immune to the rapid degradation process that was observed for the wild-type protein in iron-replete cells. Since the IRP2-specific domain was likely inserted into a duplicated gene by exon shuffling, we reasoned that it formed a separate folding domain expressed on the surface of the basic tertiary structure of IRP1. When this domain was inserted into the analogous position of IRP1, the chimeric protein acquired the capacity for iron-dependent degradation. Thus, the domain contained sufficient information to direct the rapid turnover of an associated protein (Iwai et al., 1995).

Inspection of the sequence of the iron-dependent degradation domain revealed that it contained five cysteines and a histidine, residues that could be involved in direct ligation of iron. In order to test whether cysteines were important in iron-dependent degradation, three of the cysteines of the degradation domain were simultaneously mutagenized to serines, and when this mutagenized protein was examined, we no longer observed iron-dependent degradation, thus indicating that cysteines were indispensable to iron-sensing and degradation (Iwai et al., 1995).

To better understand the degradation process, we added inhibitors of several different proteolytic systems in the cell to see which system was responsible for mediating the iron-dependent degradation. Addition of specific inhibitors of the proteasome, a large multicatalytic protease found in the cytosol of cells and thought to be responsible for the majority of protein degradation, indicated that IRP2 is degraded by the proteasome (Iwai et al., 1995; Guo et al., 1995). The ability to inhibit degradation also offered us the opportunity to characterize intermediates in the degradation pathway. We discovered that the degradation precursor of IRP2 is modified by multiple covalent additions of ubiquitin, a 76 amino acid protein that is added to proteins through the action of enzymes that hydrolyze ATP and activate ubiquitin for binding to lysine residues of target proteins. Other enzymes known as ubiquitin conjugases and ubiquitin ligases guide transfer of the ubiquitin to target proteins that presumably contain modifications that mark them for degradation. In most cases, the exact chemical nature of the protein modification that is recognized by ubiquitination enzymes is unknown.

In the case of IRP2, we had cause to believe that iron bound by the degradation domain was producing a recognizable change in the protein, either by altering conformation or by chemically modifying neighboring residues. One such chemical modification pathway, known as metal-catalyzed oxidation, is known to affect many proteins *in vitro*. The features of metal-catalyzed oxidation are that metal bound to protein reacts with oxygen, and hydroxyl radicals are generated in the vicinity of the metal binding site, leading to oxidation of neighboring residues (reviewed in Stadtman, 1993). When the side chains of amino acids such as lysines, arginines, threonines, and prolines react with hydroxyl radicals, carbonyl derivatives can be produced. As the carbonyl content of proteins can be directly measured (Levine et al., 1994), we have been able to show through use of the carbonyl assay that intermediates in the IRP2 degradation pathway are oxidized and that exposure of IRP2 to iron and oxygen generates the feature(s) recognized by ubiquitination enzymes (K. Iwai et al., in press).

In the case of each iron regulatory protein, sensing appears to depend on direct binding of iron by the protein, but the effects of iron binding on the fate of the protein are markedly different in that the outcome is degradation in the case of IRP2, whereas in IRP1, the outcome is insertion of a prosthetic group. Although the two proteins are regulated differently, high IRE binding activity in iron-depleted cells and low IRE binding activity in iron-replete cells are achieved by either form of regulation.

VI. Speculations on the Evolution of the IRE-IRP System

Characterization of the two distinct functional roles for IRP1 raises the question of how bifunctional proteins evolve. The levels of sequence homology between prokaryotic aconitases, eukaryotic mitochondria aconitases, and IRPs are shown in Table I; these homologies demonstrate that cytosolic aconitase (IRP) maintains its closest evolutionary relationship with a presumed precursor prokaryotic citric acid cycle enzyme, aconitase, whereas the mitochondrial form of the citric acid cycle protein has undergone significant deletions and sequence changes relative to the prokaryotic precursor. It is likely that IRP2 arose due to the insertion of a new exon into a duplicated IRP1 gene (Iwai et al., 1995). Although prokaryotic aconitase is closely related in sequence to IRP1, we have no evidence that any form of prokaryotic aconitase is capable of binding an IRE-RNA stem loop structure. Thus, it appears that IRP1, without losing its enzyme function, developed the ability to bind IREs.

The lability of the aconitase iron–sulfur cluster provides a means for the aconitase to be in either of two structural forms depending upon the iron status of the cell. Even in prokaryotes in which the aconitase is thought to be unifunctional, the iron–sulfur cluster of aconitase is labile and is sensitive to both oxidative stress and iron levels. Indeed, it has been proposed that the degradation of the aconitase iron–sulfur cluster in the presence of oxidative stress serves as a circuit breaker, effectively shutting off the supply of substrate for oxidative phosphorylation and thus reducing the electron and radical leakage that is characteristic of oxidative phosphorylation (Gardner and Fridovich, 1991). Thus, the presence of the apo-aconitase in cells is a consequence of the lability of iron–sulfur clusters, and accumulation of significant amounts of the apo-aconitase would be expected when there is insufficient iron available to rebuild the iron–sulfur cluster. The ability of a

TABLE I.
LEVELS OF SEQUENCE HOMOLOGY

	E. coli aconitase	IRP1
IRP1	51%	100%
IRP2	42%	58%
Yeast mitochondrial aconitase	24%	25%
Porcine mitochondrial aconitase	26%	23%

small RNA structure to specifically recognize the apo-aconitase cleft would then allow this protein transformation to be coupled to an RNA-binding event. We speculate that evolution of function may have taken place in mammalian transcripts, some of which may have acquired sequence features that allow binding to the potential RNA binding site unique to the active site region of apo-aconitase. In essence, we imagine a biological SELEX procedure (Irvine et al., 1991) in which a physiologically altered protein structure space has been sampled over time by structurally complex small regions of RNA within transcripts. The resultant binding of certain transcripts under conditions of iron depletion could produce desirable regulatory events—the potential benefits to the iron-depleted cell of decreasing ferritin biosynthesis and increasing biosynthesis of the transferrin receptor are clear. Thus, evolution would not result in acquisition of RNA binding sites in iron regulatory proteins, but rather in selection of particular RNA sequences that bind with high affinity to the unique protein sequence space of IRP1 apoprotein, a scenario that has been recently rediscovered in the development of *in vitro* RNA selection schemes. With this model in mind, one cannot help but wonder whether the IRE-IRP system is unique. It is theoretically possible to imagine an RNA-binding structure for many specific protein structures that could arise as the result of conformational changes, alterations in prosthetic groups, or covalent post-translational modifications. Such reversible modifications are characteristic of cytosolic regulation. The IRE system provides a model in which such cytosolic protein changes can be directly coupled to gene expression changes at the level of the fate of cytosolic mRNA. It is clear that such a system has been extremely successful in coordinating the regulation of multiple genes involved in the uptake, detoxification, and utilization of iron. The challenge that this system provides to biological scientists is to determine whether these sorts of RNA protein interactions, making use of the enormous potential of structural information contained within mRNA transcripts, are more widely used in RNA-based regulatory networks.

ACKNOWLEDGMENTS

We thank all our colleagues who contributed to this work. We thank Joe Harford for his significant contributions, and for creating Figure 1.

REFERENCES

Basilion, J.P., Rouault, T.A., Massinople, C.M., Klausner, R.D., and Burgess, W.H. (1994). *Proc. Natl. Acad. Sci. U.S.A.* **91,** 574–578.

Beinert, H., and Kiley, P. (1996). *FEBS Lett.* **382,** 218–219; discussion 220–221.

Binder, R., Horowitz, J.A., Basilion, J.P., Koeller, D.M., Klausner, R.D., and Harford, J.B. (1994). *EMBO J.* **13,** 1969–1980.

Casey, J.L., Hentze, M.W., Koeller, D.M., Caughman, S.W., Rouault, T.A., Klausner, R.D., and Harford, J.B. (1988). *Science* **240,** 924–928.

Casey, J.L., Koeller, D.M., Ramin, V.C., Klausner, R.D., and Harford, J.B. (1989). *EMBO J.* **8,** 3693–3699.

Caughman, S.W., Hentze, M.W., Rouault, T.A., Harford, J.B., and Klausner, R.D. (1988). *J. Biol. Chem.* **263,** 19048–19052.

Eide, D., and Guerinot, M.L., (1996). *Am. Soc. Microbiol. News* **63,** 199–205.

Gardner, P.R., and Fridovich, I. (1991). *J. Biol. Chem.* **266,** 1478–1483.

Gray, N.K., and Hentze, M.W. (1994). *EMBO J.* **13,** 3882–3891.

Guo, B., Yu, Y., and Leibold, E.A. (1994). *J. Biol. Chem.* **269,** 24252–24260.

Guo, B., Phillips, J.D., Yu, Y., and Leibold, E.A. (1995). *J. Biol. Chem.* **270,** 21645–21651.

Gutteridge, J.M. (1994). *Ann. N. Y. Acad. Sci.* **738,** 201–213.

Haile, D.J., Rouault, T.A., Harford, J.B., Kennedy, M.C., Blondin, G.A., Beinert, H., and Klausner, R.D. (1992a). *Proc. Natl. Acad. Sci. U.S.A.* **89,** 11735–11739.

Haile, D.J., Rouault, T.A., Tang, C.K., Chin, J., Harford, J.B., and Klausner, R.D. (1992b). *Proc. Natl. Acad. Sci. U.S.A.* **89,** 7536–7540.

Hentze, M.W., Caughman, S.W., Rouault, T.A., Barriocanal, J.G., Dancis, A., Harford, J.B., and Klausner, R.D. (1987a). *Science* **238,** 1570–1573.

Hentze, M.W., Rouault, T.A., Caughman, S.W., Dancis, A., Harford, J.B., and Klausner, R.D. (1987b). *Proc. Natl. Acad. Sci. U.S.A.* **84,** 6730–6734.

Irvine, D., Tuerk, C., and Gold, L. (1991). *J. Mol. Biol.* **222,** 739–761.

Iwai, K., Klausner, R.D., and Rouault, T.A. (1995). *EMBO J.* **14,** 5350–5357.

Iwai, K., Drake, S.K., Wehr, N.B., Weissman, A.B., LaVaute, T., Minato, M., Klausner, R.D., Levine, R.L., and Rouault, T.A. Oxidation, ubiquitination and degradation of IRP2; implications for degradation of other proteins. *Proc. Nat. Acad. Sci.,* in press.

Kennedy, M.C., Mende-Mueller, L., Blondin, G.A., and Beinert, H. (1992). *Proc. Natl. Acad. Sci. U.S.A.* **89,** 11730–11734.

Klausner, R.D., and Rouault, T.A. (1993). *Mol. Biol. Cell* **4,** 1–5.

Klausner, R.D., Rouault, T.A., and Harford, J.B. (1993). *Cell* **72,** 19–28.

Klausner, R.D., Ashwell, G., van Renswoude, J., Harford, J.B., and Bridges, K.R. (1983). *Proc. Natl. Acad. Sci. U.S.A.* **80,** 2263–2266.

Leibold, E.A., and Munro, H.N. (1988). *Proc. Natl. Acad. Sci. U.S.A.* **85,** 2171–2175.

Levine, R.L., Williams, J.A., Stadtman, E.R., and Shacter, E. (1994). *Methods Enzymol.* **233,** 346–357.

Maden, B.E.H. (1995). *TIBS* **20,** 337–341.

Mattia, E., Josic, D., Ashwell, G., Klausner, R., and van Renswoude, J. (1986). *J. Biol. Chem.* **261,** 4587–4593.

Mattia, E., Rao, K., Shapiro, D.S., Sussman, H.H., and Klausner, R.D. (1984). *J. Biol. Chem.* **259,** 2689–2692.

Mullner, E.W., Neupert, B., and Kuhn, L.C. (1989). *Cell* **58,** 373–382.

Munro, H.N., and Eisenstein, R.S. (1989). *Curr. Opin. Cell. Biol.* **1,** 1154–1159.

Neupert, B., Menotti, E., and Kuhn, L.C. (1995). *Nucleic Acids Res.* **23,** 2579–2583.

Philpott, C.C., Haile, D.J., Rouault, T.A., and Klausner, R.D. (1993). *J. Biol. Chem.* **268,** 17655–17658.

Philpott, C.C., Klausner, R.D., and Rouault, T.A. (1994). *Proc. Natl. Acad. Sci. U.S.A.* **91,** 7321–7325.

Rouault, T.A., Hentze, M.W., Caughman, S.W., Harford, J.B., and Klausner, R.D. (1988). *Science* **241,** 1207–1210.

Rouault, T.A., Hentze, M.W., Haile, D.J., Harford, J.B., and Klausner, R.D. (1989). *Proc. Natl. Acad. Sci. U.S.A.* **86,** 5768–5772.

Rouault, T.A., Tang, C.K., Kaptain, S., Burgess, W.H., Haile, D.J., Samaniego, F., McBride, O.W., Harford, J.B., and Klausner, R.D. (1990). *Proc. Natl. Acad. Sci. U.S.A.* **87,** 7958–7962.

Rouault, T.A., Stout, C.D., Kaptain, S., Harford, J.B., and Klausner, R.D. (1991). *Cell* **64,** 881–883.

Rouault, T.A., Haile, D.H., Downey, W.E., Philpott, C.C., Tang, C., Samaniego, F., Chin, J., Paul, I., Orloff, D., Harford, J.B., and Klausner, R. D. (1992). *BioMetals* **5,** 131–140.

Rouault, T.A., and Klausner, R.D. (1996). *Trends Biochem. Sci.* **21,** 174–177.

Rouault, T.A., Hentze, M.W., Dancis, A., Caughman, W., Harford, J.B., and Klausner, R.D. (1987). *Proc. Natl. Acad. Sci. U.S.A.* **84,** 6335–6339.

Samaniego, F., Chin, J., Iwai, K., Rouault, T.A., and Klausner, R.D. (1994). *J. Biol. Chem.* **269,** 30904–30910.

Stadtman, E.R. (1993). *Annu. Rev. Biochem.* **62,** 797–821.

Swenson, G.R., and Walden, W.E. (1994). *Nucleic Acids Res.* **22,** 2627–2633.

Tang, C.K., Chin, J., Harford, J.B., Klausner, R.D., and Rouault, T.A. (1992). *J. Biol. Chem.* **267,** 24466–24470.

Wachtershauser, G. (1992). *Prog. Biophys. Molec. Biol.* **58,** 85–201.

Weinberg, E.D. (1993). *Am. Soc. Microbiol. News* **59,** 559–562.

IMMUNITY AND GENOMICS: ANALYZING BIOLOGICAL COMPLEXITY

LEROY HOOD*

Department of Molecular Biotechnology,
University of Washington School of Medicine, Seattle, Washington

*At the time of publication, Dr. Hood's lecture manuscript had not been received.

The Harvey Lectures, Series 92, page 113

FORMER OFFICERS OF THE HARVEY SOCIETY

1905–1906

President: GRAHAM LUSK
Vice-President: SIMON FLEXNER
Treasurer: FREDERIC S. LEE
Secretary: GEORGE B. WALLACE

Council:
C.A. HERTER
S.J. MELTZER
EDWARD K. DUNHAM

1906–1907

President: GRAHAM LUSK
Vice-President: SIMON FLEXNER
Treasurer: FREDERIC S. LEE
Secretary: GEORGE B. WALLACE

Council:
C.A. HERTER
S.J. MELTZER
JAMES EWING

1907–1908

President: GRAHAM LUSK
Vice-President: JAMES EWING
Treasurer: EDWARD K. DUNHAM
Secretary: GEORGE B. WALLACE

Council:
SIMON FLEXNER
THEO. C. JANEWAY
PHILIP H. HISS, JR.

1908–1909

President: JAMES EWING
Vice-President: SIMON FLEXNER
Treasurer: EDWARD K. DUNHAM
Secretary: FRANCIS C. WOOD

Council:
GRAHAM LUSK
S.J. MELTZER
ADOLPH MEYER

1909–1910*

President: JAMES EWING
Vice-President: THEO C. JANEWAY
Treasurer: EDWARD K. DUNHAM
Secretary: FRANCIS C. WOOD

Council:
GRAHAM LUSK
S.J. MELTZER
W.J. GIES

1910–1911

President: SIMON FLEXNER
Vice-President: JOHN HOWLAND
Treasurer: EDWARD K. DUNHAM
Secretary: HAVEN EMERSON

Council:
GRAHAM LUSK
S.J. MELTZER
JAMES EWING

* At the Annual Meeting of May 18, 1909, these officers were elected. In publishing the 1909–1910 volume their names were omitted, possibly because in that volume the custom of publishing the names of the incumbents of the current year was changed to publishing the names of the officers selected for the ensuing year.

1911–1912

President: S.J. MELTZER
Vice-President: FREDERIC S. LEE
Treasurer: EDWARD K. DUNHAM
Secretary: HAVEN EMERSON

Council:
GRAHAM LUSK
JAMES EWING
SIMON FLEXNER

1912–1913

President: FREDERIC S. LEE
Vice-President: WM. H. PARK
Treasurer: EDWARD K. DUNHAM
Secretary: HAVEN EMERSON

Council:
GRAHAM LUSK
S.J. MELTZER
WM. G. MACCALLUM

1913–1914

President: FREDERIC S. LEE
Vice-President: WM. G. MACCALLUM
Treasurer: EDWARD K. DUNHAM
Secretary: AUGUSTUS B. WADSWORTH

Council:
GRAHAM LUSK
WM H. PARK
GEORGE B. WALLACE

1914–1915

President: WM. G. MACCALLUM
Vice-President: RUFUS I. COLE
Treasurer: EDWARD K. DUNHAM
Secretary: JOHN A. MANDEL

Council:
GRAHAM LUSK
FREDERIC S. LEE
W.T. LONGCOPE

1915–1916*

President: GEORGE B. WALLACE
Treasurer: EDWARD K. DUNHAM
Secretary: ROBERT A. LAMBERT

Council:
GRAHAM LUSK
RUFUS I. COLE
NELLIS B. FOSTER

1916–1917

President: GEORGE B. WALLACE
Vice-President: RUFUS I. COLE
Treasurer: EDWARD K. DUNHAM
Secretary: ROBERT A. LAMBERT

Council:
GRAHAM LUSK†
W.T. LONGCOPE
S.R. BENEDICT
HANS ZINSSER

* Dr. William G. MacCallum resigned after election. On Doctor Lusk's motion Doctor George B. Wallace was made President—no Vice President was appointed.

†Doctor Lusk was made Honorary permanent Counsellor.

1917–1918

President: EDWARD K. DUNHAM
Vice-President: RUFUS I. COLE
Treasurer: F.H. PIKE
Secretary: A.M. PAPPENHEIMER

Council:
GRAHAM LUSK
GEORGE B. WALLACE
FREDERIC S. LEE
PEYTON ROUS

1918–1919

President: GRAHAM LUSK
Vice-President: RUFUS I. COLE
Treasurer: F.H. PIKE
Secretary: K.M. VOGEL

Council:
GRAHAM LUSK
JAMES W. JOBLING
FREDERIC S. LEE
JOHN AUER

1919–1920

President: WARFIELD T. LONGCOPE
Vice-President: S.R. BENEDICT
Treasurer: F.H. PIKE
Secretary: K.M. VOGEL

Council:
GRAHAM LUSK
HANS ZINSSER
FREDERIC S. LEE
GEORGE B. WALLACE

1920–1921*

President: WARFIELD T. LONGCOPE
Vice-President: S.R. BENEDICT
Treasurer: A.M. PAPPENHEIMER
Secretary: HOMER F. SWIFT

Council:
GRAHAM LUSK
FREDERIC S. LEE
HANS ZINSSER
GEORGE B. WALLACE

1921–1922

President: RUFUS I. COLE
Vice-President: S.R. BENEDICT
Treasurer: A.M. PAPPENHEIMER
Secretary: HOMER F. SWIFT

Council:
GRAHAM LUSK
HANS ZINSSER
H.C. JACKSON
W.T. LONGCOPE

1922–1923

President: RUFUS I. COLE
Vice-President: HANS ZINSSER
Treasurer: CHARLES C. LIEB
Secretary: HOMER F. SWIFT

Council:
GRAHAM LUSK
W.T. LONGCOPE
H.C. JACKSON
S.R. BENEDICT

*These officers were elected at the Annual Meeting of May 21, 1920 but were omitted in the publication of the 1919–20 volume.

1923–1924

President: EUGENE F. DUBOIS
Vice-President: HOMER F. SWIFT
Treasurer: CHARLES C. LIEB
Secretary: GEORGE M. MACKENZIE

Council:
GRAHAM LUSK
ALPHONSE R. DOCHEZ
DAVID MARINE
PEYTON ROUS

1924–1925

President: EUGENE F. DUBOIS
Vice-President: PEYTON ROUS
Treasurer: CHARLES C. LIEB
Secretary: GEORGE M. MACKENZIE

Council:
GRAHAM LUSK
RUFUS COLE
HAVEN EMERSON
WM. H. PARK

1925–1926

President: HOMER F. SWIFT
Vice-President: H.B. WILLIAMS
Treasurer: HAVEN EMERSON
Secretary: GEORGE M. MACKENZIE

Council:
GRAHAM LUSK
EUGENE F. DUBOIS
WALTER W. PALMER
H.D. SENIOR

1926–1927

President: WALTER W. PALMER
Vice-President: WM. H. PARK
Treasurer: HAVEN EMERSON
Secretary: GEORGE M. MACKENZIE

Council:
GRAHAM LUSK
HOMER F. SWIFT
A.R. DOCHEZ
ROBERT CHAMBERS

1927–1928

President: DONALD D. VAN SLYKE
Vice-President: JAMES W. JOBLING
Treasurer: HAVEN EMERSON
Secretary: CARL A.L. BINGER

Council:
GRAHAM LUSK
RUSSEL L. CECIL
WARD J. MACNEAL
DAVID MARINE

1928–1929

President: PEYTON ROUS
Vice-President: HORATIO B. WILLIAMS
Treasurer: HAVEN EMERSON
Secretary: PHILIP D. MCMASTER

Council:
GRAHAM LUSK
ROBERT CHAMBERS
ALFRED F. HESS
H.D. SENIOR

1929–1930

President: G. CANBY ROBINSON
Vice-President: ALFRED F. HESS
Treasurer: HAVEN EMERSON
Secretary: DAYTON J. EDWARDS

Council:
GRAHAM LUSK
ALFRED E. COHN
A.M. PAPPENHEIMER
H.D. SENIOR

1930–1931

President: ALFRED E. COHN
Vice-President: J.G. HOPKINS
Treasurer: HAVEN EMERSON
Secretary: DAYTON J. EDWARDS

Council:
GRAHAM LUSK
O.T. AVERY
A.M PAPPENHEIMER
S.R. DETWILER

1931–1932

President: J.W. JOBLING
Vice-President: HOMER W. SMITH
Treasurer: HAVEN EMERSON
Secretary: DAYTON J. EDWARDS

Council:
GRAHAM LUSK
S.R. DETWILER
THOMAS M. RIVERS
RANDOLPH WEST

1932–1933

President: ALFRED F. HESS
Vice-President: HAVEN EMERSON
Treasurer: THOMAS M. RIVERS
Secretary: EDGAR STILLMAN

Council:
GRAHAM LUSK
HANS T. CLARKE
WALTER W. PALMER
HOMER W. SMITH

1933–1934

President: ALFRED HESS*
Vice-President: ROBERT K. CANNAN
Treasurer: THOMAS M. RIVERS
Secretary: EDGAR STILLMAN

Council:
STANLEY R. BENEDICT
ROBERT F. LOEB
WADE H. BROWN

1934–1935

President: ROBERT K. CANNAN
Vice-President: EUGENE L. OPIE
Treasurer: THOMAS M. RIVERS
Secretary: RANDOLPH H. WEST

Council:
HERBERT S. GASSER
B.S. OPPENHEIMER
PHILIP E. SMITH

1935–1936

President: ROBERT K. CANNAN
Vice-President: EUGENE L. OPIE
Treasurer: THOMAS M. RIVERS
Secretary: RANDOLPH H. WEST

Council:
ROBERT F. LOEB
HOMER W. SMITH
DAVID MARINE

1936–1937

President: EUGENE L. OPIE
Vice-President: PHILIP E. SMITH
Treasurer: THOMAS M. RIVERS
Secretary: MCKEEN CATTELL

Council:
GEORGE B. WALLACE
MARTIN H. DAWSON
JAMES B. MURPHY

* Dr. Hess died December 5, 1933.

1937–1938

President: EUGENE L. OPIE
Vice-President: PHILIP E. SMITH
Treasurer: THOMAS M. RIVERS
Secretary: MCKEEN CATTELL

Council:
GEORGE B. WALLACE
MARTIN H. DAWSON
HERBERT S. GASSER

1938–1939

President: PHILIP E. SMITH
Vice-President: HERBERT S. GASSER
Treasurer: KENNETH GOODNER
Secretary: MCKEEN CATTELL

Council:
HANS T. CLARKE
JAMES D. HARDY
WILLIAM S. TILLETT

1939–1940

President: PHILIP E. SMITH
Vice-President: HERBERT S. GASSER
Treasurer: KENNETH GOODNER
Secretary: THOMAS FRANCIS, JR.

Council:
HANS T. CLARKE
N. CHANDLER FOOT
WILLIAM S. TILLETT

1940–1941

President: HERBERT S. GASSER
Vice-President: HOMER W. SMITH
Treasurer: KENNETH GOODNER
Secretary: THOMAS FRANCIS, JR.

Council:
N. CHANDLER FOOT
VINCENT DU VIGNEAUD
MICHAEL HEIDELBERGER

1941–1942

President: HERBERT S. GASSER
Vice-President: HOMER W. SMITH
Treasurer: KENNETH GOODNER
Secretary: JOSEPH C. HINSEY

Council:
HARRY S. MUSTARD
HAROLD G. WOLFF
MICHAEL HEIDELBERGER

1942–1943

President: HANS T. CLARKE
Vice-President: THOMAS M. RIVERS
Treasurer: KENNETH GOODNER
Secretary: JOSEPH C. HINSEY

Council:
ROBERT F. LOEB
HAROLD G. WOLFF
WILLIAM C. VON GLAHN

1943–1944

President: HANS T. CLARKE
Vice-President: THOMAS M. RIVERS
Treasurer: COLIN M. MACLEOD
Secretary: JOSEPH C. HINSEY

Council:
ROBERT F. LOEB
WILLIAM C. VON GLAHN
WADE W. OLIVER

1944–1945

President: ROBERT CHAMBERS
Vice-President: VINCENT DU VIGNEAUD
Treasurer: COLIN M. MACLEOD
Secretary: JOSEPH C. HINSEY

Council:
WADE W. OLIVER
MICHAEL HEIDELBERGER
PHILIP D. MCMASTER

1945–1946

President: Robert Chambers
Vice-President: Vincent duVigneaud
Treasurer: Colin M. MacLeod
Secretary: Edgar G. Miller, Jr.

Council:
Philip D. McMaster
Earl T. Engle
Fred W. Stewart

1946–1947

President: Vindent duVigneaud
Vice-President: Wade W. Oliver
Treasurer: Colin M. MacLeod
Secretary: Edgar G. Miller, Jr.

Council:
Earl T. Engle
Harold G. Wolff
L. Emmett Holt, Jr.

1947–1948

President: Vincent du Vigneaud
Vice-President: Wade W. Oliver
Treasurer: Harry B. Van Dyke
Secretary: Maclyn McCarty

Council:
Paul Klemperer
L. Emmett Holt, Jr.
Harold G. Wolff

1948–1949

President: Wade W. Oliver
Vice-President: Robert F. Loeb
Treasurer: Harry B. Van Dyke
Secretary: Maclyn McCarty

Council:
Paul Klemperer
Severo Ochoa
Harold L. Temple

1949–1950

President: Wade W. Oliver
Vice-President: Robert F. Loeb
Treasurer: James B. Hamilton
Secretary: Maclyn McCarty

Council:
William S. Tillett
Severo Ochoa
Harold L. Temple

1950–1951

President: Robert F. Loeb
Vice-President: Michael Heidelberger
Treasurer: James B. Hamilton
Secretary: Ludwin W. Eichna

Council:
William S. Tillett
A.M. Pappenheimer, Jr.
David P. Barr

1951–1952

President: René J. Dubois
Vice-President: Michael Heidelberger
Treasurer: James B. Hamilton
Secretary: Ludwin W. Eichna

Council:
David P. Barr
Robert F. Pitts
A.M. Pappenheimer, Jr.

1952–1953

President: Michael Heidelberger
Vice-President: Severo Ochoa
Treasurer: Chandler McC. Brooks
Secretary: Henry D. Lauson

Council:
Robert F. Pitts
Jean Oliver
Alexander B. Gutman

1953–1954

President: SEVERO OCHOA
Vice-President: DAVID P. BARR
Treasurer: CHANDLER McC. BROOKS
Secretary: HENRY D. LAUSON

Council:
JEAN OLIVER
ALEXANDER B. GUTMAN
ROLLIN D. HOTCHKISS

1954–1955

President: DAVID P. BARR
Vice-President: COLIN M. MACLEOD
Treasurer: CHANDLER McC. BROOKS
Secretary: HENRY D. LAUSON

Council:
ALEXANDER B. GUTMAN
ROLLIN D. HOTCHKISS
DAVID SHEMIN

1955–1956

President: COLIN M. MACLEOD
Vice-President: FRANK L. HORSFALL, JR.
Treasurer: CHANDLER McC. BROOKS
Secretary: RULON W. RAWSON

Council:
ROLLIN D. HOTCHKISS
DAVID SHEMIN
ROBERT F. WATSON

1956–1957

President: FRANK L. HORSFALL, JR.
Vice-President: WILLIAM S. TILLETT
Treasurer: CHANDLER McC. BROOKS
Secretary: RULON W. RAWSON

Council:
DAVID SHEMIN
ROBERT F. WATSON
ABRAHAM WHITE

1957–1958

President: WILLIAM S. TILLETT
Vice-President: ROLLIN D. HOTCHKISS
Treasurer: CHANDLER McC. BROOKS
Secretary: H. SHERWOOD LAWRENCE

Council:
ROBERT F. WATSON
ABRAHAM WHITE
JOHN V. TAGGART

1958–1959

President: ROLLIN D. HOTCHKISS
Vice-President: ANDRE COURNAND
Treasurer: CHANDLER McC. BROOKS
Secretary: H. SHERWOOD LAWRENCE

Council:
ABRAHAM WHITE
JOHN V. TAGGART
WALSH McDERMOTT

1959–1960

President: ANDRE COURNAND
Vice-President: ROBERT F. PITTS
Treasurer: EDWARD J. HEHRE
Secretary: H. SHERWOOD LAWRENCE

Council:
JOHN V. TAGGART
WALSH McDERMOTT
ROBERT F. FURCHGOTT

1960–1961

President: ROBERT F. PITTS
Vice-President: DICKINSON W. RICHARDS
Treasurer: EDWARD J. HEHRE
Secretary: ALEXANDER G. BEARN

Council:
WALSH McDERMOTT
ROBERT F. FURCHGOTT
LUDWIG W. EICHNA

1961–1962

President: DICKINSON W. RICHARDS
Vice-President: PAUL WEISS
Treasurer: I. HERBERT SCHEINBERG
Secretary: ALEXANDER G. BEARN

Council:
ROBERT F. FURCHGOTT
LUDWIG W. EICHNA
EFRAIM RACKER

1962–1963

President: PAUL WEISS
Vice-President: ALEXANDER B. GUTMAN
Treasurer: I. HERBERT SCHEINBERG
Secretary: ALEXANDER G. BEARN

Council:
LUDWIG W. EICHNA
EFRAIM RACKER
ROGER L. GREIF

1963–1964

President: ALEXANDER B. GUTMAN
Vice-President: EDWARD L. TATUM
Treasurer: SAUL J. FARBER
Secretary: ALEXANDER G. BEARN

Council:
EFRAIM RACKER
ROGER L. GREIF
IRAVING M. LONDON

1964–1965

President: EDWARD TATUM
Vice-President: CHANDLER McC. BROOKS
Treasurer: SAUL J. FARBER
Secretary: RALPH L. ENGLE, JR.

Council:
ROGER L. GREIF
LEWIS THOMAS
IRVING M. LONDON

1965–1966

President: CHANDLER McC. BROOKS
Vice-President: ABRAHAM WHITE
Treasurer: SAUL J. FARBER
Secretary: RALPH L. ENGLE, JR.

Council:
IRVING M. LONDON
LEWIS THOMAS
GEORGE K. HIRST

1966–1967

President: ABRAHAM WHITE
Vice-President: RACHMIEL LEVINE
Treasurer: SAUL J. FARBER
Secretary: RALPH L. ENGLE, JR.

Council:
LEWIS THOMAS
GEORGE K. HIRST
DAVID NACHMANSOHN

1967–1968

President: RACHMIEL LEVINE
Vice-President: SAUL J. FARBER
Treasurer: PAUL A. MARKS
Secretary: RALPH L. ENGLE, JR.

Council:
GEORGE K. HIRST
DAVID NACHMANSOHN
MARTIN SONENBERG

1968–1969

President: SAUL J. FARBER
Vice-President: JOHN V. TAGGART
Treasurer: PAUL A. MARKS
Secretary: ELLIOTT F. OSSERMAN

Council:
DAVID NACHMANSOHN
MARTIN SONENBERG
HOWARD A. EDER

1969–1970

President: JOHN V. TAGGART
Vice-President: BERNARD L. HORECKER
Treasurer: PAUL A. MARKS
Secretary: ELLIOTT F. OSSERMAN

Council:
MARTIN SONENBERG
HOWARD A. EDER
SAUL J. FARBER

1970–1971

President: BERNARD L. HORECKER
Vice-President: MACLYN MCCARTY
Treasurer: EDWARD C. FRANKLIN
Secretary: ELLIOTT F. OSSERMAN

Council:
HOWARD A. EDER
SAUL J. FARBER
SOLOMON A. BERSON

1971–1972

President: MACLYN MCCARTY
Vice-President: ALEXANDER G. BEARN
Treasurer: EDWARD C. FRANKLIN
Secretary: ELLIOTT F. OSSERMAN

Council:
SAUL J. FARBER
SOLOMON A. BERSON
HARRY EAGLE

1972–1973

President: ALEXANDER G. BEARN
Vice-President: PAUL A. MARKS
Treasurer: EDWARD C. FRANKLIN
Secretary: JOHN ZABRISKIE

Council:
HARRY EAGLE
JERARD HURWITZ

1973–1974

President: PAUL A. MARKS
Vice-President: IGOR TAMM
Treasurer: EDWARD C. FRANKLIN
Secretary: JOHN B. ZABRISKIE

Council:
HARRY EAGLE
CHARLOTTE FRIEND
JERARD HURWITZ

1974–1975

President: IGOR TAMM
Vice-President: GERALD M. EDELMAN
Treasurer: STEPHEN I. MORSE
Secretary: JOHN B. ZABRISKIE

Council:
JERARD HURWITZ
H. SHERWOOD LAWRENCE
CHARLOTTE FRIEND

1975–1976

President: GERALD M. EDELMAN
Vice-President: ELVIN A. KABAT
Treasurer: STEPHEN I. MORSE
Secretary: JOHN B. ZABRISKIE

Council:
PAUL A. MARKS
H. SHERWOOD LAWRENCE
CHARLOTTE FRIEND

1976–1977

President: ELVIN A. KABAT
Vice-President: FRED PLUM
Treasurer: STEPHEN I. MORSE
Secretary: DONALD M. MARCUS

Council:
H. SHERWOOD LAWRENCE
PAUL A. MARKS
BRUCE CUNNINGHAM

1977–1978

President: FRED PLUM
Vice-President: CHARLOTTE FRIEND
Treasurer: STEPHEN I. MORSE
Secretary: DONALD M. MARCUS

Council:
PAUL A. MARKS
BRUCE CUNNINGHAM
VITTORIO DEFENDI

1978–1979

President: CHARLOTTE FRIEND
Vice-President: MARTIN SONENBERG
Treasurer: ALFRED STRACHER
Secretary: DONALD M. MARCUS

Council:
BRUCE CUNNINGHAM
VITTORIO DEFENDI
DEWITT S. GOODMAN

1979–1980

President: MARTIN SONENBERG
Vice-President: KURT HIRSCHHORN
Treasurer: ALFRED STRACHER
Secretary: EMIL C. GOTSCHLICH

Council:
VITTORIO DEFENDI
DEWITT S. GOODMAN
ORA ROSEN

1980–1981

President: KURT HIRSCHHORN
Vice-President: GERALD WEISSMANN
Treasurer: ALFRED STRACHER
Secretary: EMIL C. GOTSCHLICH

Council:
RALPH NACHMAN
DEWITT S. GOODMAN
ORA ROSEN

1981–1982

President: GERALD WEISSMANN
Vice-President: DEWITT S. GOODMAN
Treasurer: ALFRED STRACHER
Secretary: EMIL C. GOTSCHLICH

Council:
KURT HIRSCHHORN
RALPH L. NACHMAN
ORA ROSEN

1982–1983

President: DEWITT S. GOODMAN
Vice-President: MATTHEW D. SCHARFF
Treasurer: ALFRED STRACHER
Secretary: EMIL C. GOTSCHLICH

Council:
KURT HIRSCHHORN
RALPH L. NACHMAN
GERALD WEISSMANN

1983–1984

President: MATTHEW D. SCHARFF
Vice-President: HAROLD S. GINSBERG
Treasurer: RICHARD A. RIFKIND
Secretary: EMIL C. GOTSCHLICH

Council:
KURT HIRSCHHORN
GERALD WEISSMANN
JAMES P. QUIGLEY

1984–1985

President: HAROLD S. GINSBERG
Vice-President: JAMES E. DARNELL
Treasurer: RICHARD A. RIFKIND
Secretary: ROBERT J. DESNICK

Council:
JAMES P. QUIGLEY
MATTHEW D. SCHARFF
GERALD WEISSMANN

1985–1986

President: JAMES E. DARNELL
Vice-President: DAVID SABATINI
Treasurer: RICHARD A. RIFKIND
Secretary: ROBERT J. DESNICK

Council:
JAMES P. QUIGLEY
MATTHEW D. SCHARFF
HAROLD S. GINSBERG

1986–1987

President: DAVID SABATINI
Vice-President: DONALD A. FISCHMAN
Treasurer: RICHARD A. RIFKIND
Secretary: ROBERT J. DESNICK

Council:
JAMES E. DARNELL
HAROLD S. GINSBERG
MATTHEW D. SCHARFF

1987–1988

President: DONALD A. FISCHMAN
Vice-President: JONATHAN R. WARNER
Treasurer: RICHARD A. RIFKIND
Secretary: ROBERT J. DESNICK

Council:
JAMES E. DARNELL
HAROLD S. GINSBERG
WILLIAM MCALLISTER
DAVID SABATINI

1988–1989

President: JONATHAN R. WARNER
Vice-President: ISIDORE S. EDELMAN
Treasurer: JOSEPH R. BERTINO
Secretary: ROBERT J. DESNICK

Council:
JAMES E. DARNELL
DONALD A. FISCHMAN
WILLIAM MCALLISTER
DAVID SABATINI

1989–1990

President: ISIDORE S. EDELMAN
Vice-President: DAVID J. LUCK
Treasurer: JOSEPH R. BERTINO
Secretary: PETER PALESE

Council:
DONALD A. FISCHMAN
ROCHELLE HIRSCHHORN
WILLIAM MCALLISTER
JONATHAN R. WARNER

1990–1991

President: DAVID J. LUCK
Vice-President: KENNETH BERNS
Treasurer: JOSEPH R. BERTINO
Secretary: PETER PALESE

Council:
ISIDORE S. EDELMAN
ROCHELLE HIRSCHHORN
WILLIAM MCALLISTER
JONATHAN R. WARNER

1991–1992

President: KENNETH I. BERNS
Vice-President: JERARD HURWITZ
Treasurer: JOSEPH R. BERTINO
Secretary: PETER PALESE

Council:
ISIDORE S. EDELMAN
ROCHELLE HIRSCHHORN
SUSAN HORWITZ
M.A.Q. SIDDIQUI

1992–1993

President: JERARD HURWITZ
Vice-President: RUTH S. NUSSENZWEIG
Treasurer: JOSEPH R. BERTINO
Secretary: PETER PALESE

Council:
KENNETH I. BERNS
SUSAN HORWITZ
M.A.Q. SIDDIQUI
SAMUEL C. SILVERSTEIN

1993–1994

President: RUTH S. NUSSENZWEIG
Vice-President: ROBERT ROEDER
Treasurer: JOAN MASSAGUE
Secretary: PAUL B. LAZAROW

Council:
SUSAN HORWITZ
JERARD HURWITZ
M.A.Q. SIDDIQUI
SAMUEL C. SILVERSTEIN

1994–1995

President: ROBERT ROEDER
Vice-President: DAVID HIRSH
Treasurer: JOAN MASSAGUE
Secretary: PAUL B. LAZAROW

Council:
EVA B. CRAMER
LORRAINE GUDAS
MARSHALL HORWITZ
JERARD HURWITZ
RUTH NUSSENZWEIG
SAMUEL C. SILVERSTEIN

1995–1996

President: DAVID I. HIRSH
Vice-President: STUART AARONSON
Treasurer: SCOTT EMMONS
Secretary: PAUL B. LAZAROW

Council:
EVA B. CRAMER
LORRAINE GUDAS
MARSHALL HORWITZ
KENNETH MARIANS
LENNART PHILIPSON
ROBERT G. ROEDER

CUMULATIVE AUTHOR INDEX*

*(h), honorary; (a), active; (d), deceased.
†Did not present lecture because of World War II.

DR. E. MARGOLLASH, 1970–71 (h)
PROF A. MAGNUS-LEVY, 1909-10 (d)
DR. H.W. MAGOUN, 1951–52 (h)
PROF. PHILIP W. MAJERUS, 1986–87 (h)
DR. F.B. MALLORY, 1912–13 (d)
DR. THOMAS P. MANIATIS, 1986–87 (h)
DR. FRANK C. MANN, 1927–28 (d)
DR. DAVID MARINE, 1923–24 (d)
DR. CLEMENT L. MARKERT, 1963–64 (h)
DR. PAUL A. MARKS, 1970–71 (a)
DR. PHILIPPA MARRACK, 1993–94
DR. GUY MARRIAN, 1938–39 (h)
PROF. W. MCKIM MARRIOTT, 1919–20 (d)
DR. E.K. MARSHALL, JR., 1929–30 (d)
DR. BRIAN W. MATTHEWS, 1986–86 (h)
DR. MANFRED M. MAYER, 1976–77 (h)
DR. DANIEL MAZIA, 1957–58 (h)
DR. MACLYN MCCARTY, 1969–70 (a)
PROF. E.V. MCCOLLUM, 1916–17 (d)
DR. WALSH MCDERMOTT, 1967–68 (h)
DR. HARDEN M. MCDONNELL (h)
DR. W.D. MCELROY, 1955–56 (h)
DR. STEVEN LANIER MCKNIGHT, 1991–92 (h)
DR. PHILIP D. MCMASTER, 1941–42 (h)
DR. P.B. MEDAWAR, 1956–57 (h)
DR. WALTER J. MEEK, 1940–41 (d)
PROF. ALTON MEISTER, 1967–68 (h)
DR. JOHN J. MEKALANOS, 1993–94
DR. S.J. MELTZER, 1906–07 (d)
PROF. LAFAYETTE B. MENDEL, 1905–06, 1914–15 (d)
DR. R. BRUCE MERRIFIELD, 1971–72 (h)
DR. HENRY METZGER, 1984–85 (h)
PROF. ADOLPH MEYER, 1909–10 (d)
PROF. HANS MEYER, 1905–06 (d)
DR. KARL MEYER, 1955–56 (h)
DR. K.F. MEYER, 1939–40 (d)
DR. OTTO MEYERHOF, 1922–23 (d)
DR. LEONOR MICHAELIS, 1926–27 (d)
DR. WILLIAM S. MILLER, 1924–25 (d)
PROF. CHARLES S. MINOT, 1905–06 (d)
DR. GEORGE R. MINOT, 1927–28 (d)
DR. BEATRICE MINTZ, 1975–76 (h)
DR. A.E. MIRSKY, 1950–51 (h)
DR. JACQUES MONOD, 1961–62 (h)
DR. CARL V. MOORE, 1958–59 (h)
DR. FRANCIS D. MOORE, 1956–57 (h)
DR. STANFORD MOORE, 1956–57 (h)
PROF. T.H. MORGAN, 1905–06 (d)
DR. GIUSEPPE MORUZZI, 1962–63 (h)
DR. J. HOWARD MUELLER, 1943–44 (d)
PROF. FRIEDRICH MULLER, 1906–07 (d)
DR. H.J. MULLER, 1947–48 (d)
DR. HANS MÜLLER-EBERHARD, 1970–71 (a)
PROF. JOHN R. MURLIN, 1916–17 (d)
DR. W.P. MURPHY, 1927–28 (d)
DR. DAVID NACHMANSOHN, 1953–54 (h)
DR. F.R. NAGER, 1925–26 (d)
DR. KIM NASMYTH, 1992–93
DR. DANIEL NATHANS, 1974–75 (h)
DR. STANLEY G. NATHENSON, 1993–94 (a)
DR. JAMES V. NEEL, 1960–61 (h)
DR. ELIZABETH F. NEUFELD, 1979–80 (h)
DR. FRED NEUFELD, 1926–27 (d)
SIR ARTHUR NEWSHOLME, 1920–21 (d)
DR. MARSHALL W. NIRENBERG, 1963–64 (h)
DR. HIDEYO NOGUVHI, 1915–16 (d)
DR. HARRY F. NOLLER, 1988–89 (a)
DR. JOHN H. NORTHROP, 1925–26, 1934–35 (d)
DR. G.J.V. NOSSAL, 1967–68 (h)
PROF. FREDERICK G. NOVY, 1934–35 (d)
DR. SHOSAKU NUMA, 1987–88 (a)
DR. PAUL NURSE, 1996–97
DR. RUTH S. NUSSENZWEIG, 1982–83 (a)
DR. VICTOR NUSSENZWEIG, 1982–83 (a)
DR. CHRISTIANE NÜSSLEIN-VOLHARD, 1990–91 (h)
PROF. GEORGE H.F. NUTTALL, 1912–13 (d)
DR. SEVERO OCHOA, 1950–51 (a)
DR. LLOYD J. OLD, 1971–72, 1975–76 (h)
DR. JOHN OLIPHANT, 1943–44 (d)
DR. JEAN OLIVER, 1944–45 (h)
DR. BERT W. O'MALLEY, 1976–77 (h)
DR. J.L. ONCLEY, 1954–55 (h)
DR. EUGENE L. OPIE, 1909–10, 1928–29, 1954–55 (d)
DR. STUART H. ORKIN, 1987–88 (a)
PROF. HENRY F. OSBORN, 1911–12 (d)

DR. MARY JANE OSBORN, 1982–83 (h)
DR. THOMAS B. OSBORNE, 1910–11 (d)
DR. WINTHROP J.V. OSTERHOUT, 1921–22, 1929–30 (d)
DR. NORMAN R. PACE, 1995–96
DR. GEORGE E. PALADE, 1961–62 (a)
DR. A.M. PAPPENHEIMER, JR., 1956–57, 1980–81 (a)
DR. JOHN R. PAPPENHEIMER, 1965–66 (a)
PROF. ARTHUR B. PARDEE, 1969–70 (h)
DR. EDWARDS A. PARK, 1938–39 (d)
PROF. W.H. PARK, 1905–06 (d)
PROF. G.H. PARKER, 1913–14 (d)
DR. STEWART PATON, 1917–19 (d)
DR. JOHN R. PAUL, 1942–43 (d)
DR. L. PAULING, 1953–54 (h)
DR. FRANCIS W. PEABODY, 1916–17 (d)
PROF. RICHARD M. PEARCE, 1909–10 (d)
DR. RAYMOND PEARL, 1921–22 (d)
DR. WILLIAM STANLEY PEART, 1977–78 (h)
DR. WILDER PENFIELD, 1936–37 (d)
DR. M.F. PERUTZ, 1967–68 (h)
DR. JOHN P. PETERS, 1937–38 (d)
DR. W.H. PETERSON, 1946–47 (d)
DR. DAVID C. PHILLIPS, 1970–71 (h)
DR. ERNST P. PICK, 1929–30 (h)
DR. LUDWIG PICK, 1931–32 (d)
DR. GREGORY PINCUS, 1966–67 (d)
DR. CLEMENS PIRQUET, 1921–22 (d)
DR. COLIN PITENDRIGH, 1960–61 (h)
DR. ROBERT PITTS, 1952–53 (d)
DR. A. POLICARD, 1931–32 (h)
PROF. GEORGE J. POPJAK, 1969–70 (h)
DR. KEITH R. PORTER, 1955–56 (a)
PROF. RODNEY R. PORTER, 1969–70 (h)
DR. W.T. PORTER, 1906–07, 1917–19 (d)
DR. STANLEY B. PRUSINER, 1991–92 (h)
DR. PETER PRYCIAK, 1994–95
DR. MARK PTASHNE, 1973–74 (h)
DR. T.T. PUCK, 1958–59 (h)
DR. J.J. PUTNAM, 1911–12 (d)
DR. VINCENT R. RACANIELLO, 1991–92 (h)
DR. EFRAIM RACKER, 1955–56 (a)
DR. HERMANN RAHN, 1958–59 (h)
DR. CHARLES H. RAMMELKAMP, JR., 1955–56 (h)
DR. S. WALTER RANSON, 1937–37 (d)
DR. KENNETH B. RAPER, 1961–62 (h)
DR. ALEXANDER RICH, 1982–83 (a)
DR. ARNOLD R. RICH, 1946–57 (d)
PROF. ALFRED N. RICHARDS, 1920–21, 1934–35 (d)
DR. DICKINSON W. RICHARDS, 1943–44 (h)
PROF. THEODORE W. RICHARDS, 1911–12 (d)
DR. CURT P. RICHTER, 1942–43 (h)
DR. D. RITTENBERG, 1948–49 (d)
DR. THOMAS M. RIVERS, 1933–34 (d)
DR. WILLIAM ROBBINS, 1942–43 (h)
DR. O.H. ROBERTSON, 1943–43 (d)
PROF. WILLIAM C. ROSE, 1934–35 (h)
PROF. ORA MENDELSOHN ROSEN, 1986–87 (a)
DR. M.J. ROSENAU, 1908–09 (d)
DR. RUSSELL ROSS, 1981–82 (a)
DR. MICHAEL G. ROSSMANN, 1987–88 (a)
DR. JESSE ROTH, 1981–82 (a)
DR. JAMES E. ROTHMAN, 1990–91 (h)
DR. F.J.W. ROUGHTON, 1943–44 (h)
DR. PEYTON ROUS, 1935–36 (d)
DR. WALLACE P. ROWE, 1975–76 (h)
DR. GERALD M. RUBIN, 1987–88 (a)
DR. HARRY RUBIN, 1965–66 (h)
PROF. MAX RUBNER, 1912–13 (d)
DR. FRANK H. RUDDLE, 1973–74 (h)
DR. JOHN RUNNSTROM, 1950–51 (h)
DR. ERRKI RUOSLAHTI, 1988–89 (a)
MAJOR FREDERICK F. RUSSELL, 1912–13 (d)
DR. F.R. SABIN, 1915–16 (d)
DR. LEO SACHS, 1972–73 (h)
DR. RUTH SAGER, 1982–83 (h)
DR. BENGT SAMUELSSON, 1979–80 (h)
DR. WILBUR A. SAWYER, 1934–35 (d)
DR. HOWARD SCHACHMAN, 1972–73 (h)
PROF. E.A. SCHAFER, 1907–08 (d)
DR. GOTTFRIED SCHATZ, 1989–90 (h)
DR. ROBERT T. SCHIMKE, 1980–81 (h)
DR. MATTHEW D. SCHARFF, 1973–74 (a)
DR. HAROLD W. SCHERAGA, 1967–68 (h)
DR. BELA SCHICK, 1922–23 (h)
DR. ROBERT SCHLEIF, 1988–89 (a)

DR. JOSEPH SCHLESSINGER, 1993–94
DR. OSCAR SCHLOSS, 1924–25 (d)
DR. STUART F. SCHLOSSMAN, 1983–84 (h)
PROF. ADOLPH SCHMIDT, 1913–14 (d)
DR. CARL F. SCHMIDT, 1948–49 (h)
DR. KNUT SCHMIDT-NEILSEN, 1962–63 (h)
DR. RANDY W. SCHEKMAN, 1994–95
DR. FRANCIS O. SCHMITT, 1944–45 (h)
DR. R. SCHOENEHEIMER, 1936–37 (d)
DR. P.F. SCHOLANDER, 1961–62 (h)
DR. STUART LEE SCHREIBER, 1995–96
DR. KATHRIN SCHRICK, 1994–95
DR. ROBERT S. SCHWARTZ, 1985–86 (h)
DR. NEVIN S. SRIMSHAW, 1962–63 (h)
DR. WILLIAM H. SEBRELL, 1943–44 (h)
PROF. W.T. SEDGWICK, 1911–12 (d)
DR. WALTER SEEGERS, 1951–52 (h)
DR. J. EDWIN SEEGMILLER, 1969–70 (h)
DR. MICHAEL SELA, 1971–72 (h)
DR. PHLIP A. SHAFFER, 1922–23 (d)
DR. JAMES A. SHANNON, 1945–46 (h)
DR. LUCY SIIAPIRO, 1992–93
DR. PHILLIP A. SHARP, 1985–86 (h)
DR. DAVID SHEMIN, 1954–55 (h)
DR. HENRY C. SHERMAN, 1917–19 (d)
DR. RICHARD SHOPE, 1935–36 (d)
DR. EPHRAIM SHORR, 1954–55 (d)
DR. KAI SIMONS, 1993–94
DR. ROBERT L. SINSHEIMER, 1968–69 (h)
DR. E.C. SLATER, 1970–71 (h)
DR. G. ELIOT SMITH, 1930–31 (d)
DR. EMIL L. SMITH, 1966–67 (h)
DR. HOMER W. SMITH, 1939–40 (d)
DR. PHILIP E. SMITH, 1929–30 (d)
PROF. THEOBALD SMITH, 1905–06 (d)
DR. GEORGE D. SNELL, 1978–79 (h)
DR. SOLOMON H. SNYDER, 1977–78 (h)
DR. LOUIS SOKOLOFF, 1983–84 (h)
DR. T.M. SONNEBORN, 1948–49 (h)
DR. S.P.L. SORENSON, 1924–25 (d)
DR. CARL C. SPEIDEL, 1940–41 (h)
DR. SOL SPIEGELMAN, 1968–69 (a)
DR. ROGER W. SPERRY, 1966–67 (h)
DR. TIMOTHY A. SPRINGER, 1993–94
DR. WILLIAM C. STADIE, 1941–42 (d)
DR. EARL R. STADTMAN, 1969–70 (h)
DR. ROGER STANIER, 1959–60 (h)
DR. WENDELL STANLEY, 1937–38 (d)
DR. EARNEST H. STARLING, 1907–08 (d)
DR. ISAAC STARR, 1946–47 (h)
DR. WILLIAM H. STEIN, 1956–57 (h)
DR. DONALD F. STEINER, 1982–83 (h)
DR. JOAN A. STEITZ, 1984–85 (h)
DR. P. STETSON, 1927–28 (d)
PROF. GEORGE STEWART, 1912–13 (d)
PROF. CH. WARDELL STILES, 1915–16 (d)
DR. BRUCE STILLMAN, 1992–93
DR. C.R. STOCKARD, 1921–22 (d)
DR. WALTER STRAUB, 1928–29 (h)
DR. GEORGE L. STREETER, 1933–34 (h)
DR. JACK L. STROMINGER, 1968–69 (h)
DR. R.P. STRONG, 1913–14 (d)
DR. LUBERT STRYER, 1991–92 (h)
PROF. EARL W. SUTHERLAND, JR., 1961–62 (d)
PROF. HOMER F. SWIFT, 1919–20 (d)
DR. W.W. SWINGLE, 1931–32 (d)
DR. V.P. SYDENSTRICKER, 1942–43 (h)
DR. ALBERT SZENT-GYORGYI, 1938–39 (h)
DR. W.H. TALIAFERRO, 1931–32 (d)
PROF. ALONZO E. TAYLOR, 1907–08 (d)
DR. HOWARD M. TEMIN, 1973–74 (h)
PROF. W.S. THAYER, 1911–12 (d)
DR. HUGO THEORELL, 1965–66 (h)
DR. LEWIS THOMAS, 1967–68 (a)
DR. SHIRLEY M. TILGHMAN, 1991–92 (h)
DR. WILLIAM S. TILLETT, 1949–50 (h)
DR. ARNE TISELIUS, 1939–40 (h)
DR. ROBERT TJIAN, 1994–95
DR. A.R. TODD, 1951–52 (h)
DR. GORDON M. TOMKINS, 1972–73 (h)
DR. SUSUMU TONEGAWA, 1979–80 (h)
DR. SIDNEY UDENFRIEND, 1964–65 (h)
COLONEL F.P. UNDERHILL, 1917–19 (d)
DR. HANS USSING, 1963–64 (h)
DR. P. ROY VAGELOS, 1974–75 (h)
DR. DONALD D. VAN SLYKE, 1915–16 (d)
DR. HAROLD VARMUS, 1987–88 (a)
DR. MARTHA VAUGHAN, 1981–82 (h)
PROF. VICTOR C. VAUGHN, 1913–14 (d)
PROF. MAX VERWORN, 1911–12 (d)

PROF. CARL VOEGTLIN, 1919–20 (d)
DR. U.S. VON EULER, 1958–59 (h)
DR. ALEXANDER VON MURALT, 1947–48 (h)
PROF. CARL VON NOORDEN, 1905–06 (d)
DR. SELMAN A. WAKSMAN, 1944–45 (d)
DR. GEORGE WALD, 1945–46 (h)
DR. JAN WALDENSTROM, 1960–61 (h)
PROF. THOMAS A. WALDMANN, 1986–87 (h)
DR. PETER WALKER, 1995–96
PROF. AUGUSTUS D. WALLER, 1913–14 (d)
DR. JAMES C. WANG, 1985–86 (h)
DR. JOSEF WARKANY, 1952–53 (h)
COLONEL STAFFORD L. WARREN, 1945–46 (h)
DR. ALFRED S. WARTHIN, 1917–19 (d)
DR. C.J. WATSON, 1948–49 (h)
DR. JOSEPH T. WEARN, 1939–40 (h)
DR. H.H. WEBER, 1953–54 (d)
PROF. J. CLARENCE WEBSTER, 1905–06 (d)
DR. L.T. WEBSTER, 1931–32 (d)
DR. A. ASHLEY WEECH, 1938–39 (h)
DR. SILVIO WEIDMANN, 1965–66 (h)
DR. ROBERT A. WEINBERG, 1984–85 (h)
DR. HAROLD M. WEINTRAUB, 1983–84 (h)
DR. PAUL WEISS, 1958–59 (h)
IRVING L. WEISSMAN, 1989–90 (h)
DR. CHARLES WEISSMANN, 1981–82 (h)
DR. WILLIAM H. WELCH, 1915–16 (d)
DR. THOMAS H. WELLER, 1956–57 (h)
PROF. H. GIDEON WELLS, 1910–11 (d)
DR. K.F. WENCKEBACH, 1922–23 (d)
DR. GEORGE H. WHIPPLE, 1921–22 (d)
DR. ABRAHAM WHITE, 1947–48 (a)
DR. RAYMOND L. WHITE, 1984–85 (h)
DR. CARL J. WIGGERS, 1920–21, 1956–57 (d)
DR. V.B. WIGGLESWORTH, 1959–60 (h)
DR. DON C. WILEY, 1988–89 (a)
DR. CARROLL M. WILLIAMS, 1951–52 (h)
DR. LINSLEY R. WILLIAMS, 1917–19 (d)
DR. RICHARD WILLSTATTER, 1926–27 (d)
DR. EDMUND B. WILSON, 1906–07 (d)
DR. EDWIN B. WILSON, 1925–26 (d)
PROF. J. GORDON WILSON, 1917–19 (h)
DR. JEAN D. WILSON, 1983–84 (h)
DR. WILLIAM F. WINDLE, 1944–45 (h)
DR. F.R. WINTON, 1951–52 (h)
DR. MAXWELL M. WINTROBE, 1949–50 (d)
PROF. S.B. WOLFBACH, 1920–21 (d)
DR. HAROLD G. WOLFF, 1943–44 (d)
DR. HARLAND G. WOOD, 1949–50 (h)
DR. W. BARRY WOOD, JR., 1951–52 (d)
DR. WILLIAM B. WOOD, 1977–78 (h)
PROF. SIR MICHAEL F.A. WOODRUFF, 1970–71 (h)
DR. ROBERT B. WOODWARD, 1963–64 (h)
DR. R.T. WOODYATT, 1915–16 (d)
DR. D.W. WOOLLEY, 1945–46 (d)
SIR ALMROTH E. WRIGHT, 1906–07 (d)
DR. ROSALYN S. YALOW, 1966–67 (h)
DR. KEITH R. YAMAMOTO, 1995–96
PROF. ROBERT M. YERKES, 1917–19, 1935–36 (d)
DR. PAUL C. ZAMECNIK, 1959–60 (h)
DR. L. ZECHMEISTER, 1951–52 (h)
DR. NORTON D. ZINDER, 1966–67 (a)
DR. R.M. ZINKERNAGEL, 1993–94
PROF. HANS ZINSSER, 1914–15 (d)

ACTIVE MEMBERS

Dr. Bent Aasted
Dr. Ruth Abramson
Dr. Alan Aderem
Dr. Salah Al-Askari
Dr. Qais Al-Awqati
Dr. Emma G. Allen
Dr. Norman R. Alpert
Dr. Blanche Pearl Alter
Dr. Burton M. Altura
Dr. Efrain Charles Azmitia
Dr. Richard A. Bader
Dr. Silvio Baez
Dr. David S. Baldwin
Dr. M. Earl Balis
Dr. Amiya Banerjee
Dr. Arthur Bank
Dr. James J. Barondess
Dr. Robert Bases
Dr. Claudio Basilico
Dr. Hagan Bayley
Dr. A. Robert Beck
Dr. Carl G. Becker
Dr. Joseph W. Becker
Dr. John Michael Bedford
Dr. Baruj Benacerraf
Dr. Robert Benezra
Dr. Craig J. Benham
Dr. Thomas P. Bennett
Dr. Gordon D. Benson
Dr. Carole L. Berger
Dr. Lawrence Berger
Dr. Gregory I. Berk
Dr. Paul D. Berk
Dr. Kenneth I. Berns
Dr. Carl A. Berntsen
Dr. Joseph R. Bertino
Dr. Rajesh M. Bhatnagar
Dr. R.J. Bing*
Dr. David Bishop
Dr. John Anthony Blaho
Dr. Beverly F. Blatt
Dr. Andrew Blitzer
Dr. Carl P. Blobel
Dr. Gunter K.J. Blobel
Dr. Konrad Bloch
Dr. Richard Steven Bockman
Dr. Adele L. Boskey
Dr. Robert J. Boylan
Dr. Thomas B. Bradley
Dr. Leon Bradlow
Dr. Goodwin M. Breinin
Dr. Patrick J. Brennwald
Dr. Esther Breslow
Dr. Robin W. Briehl
Dr. Felix Bronner
Dr. Mark S. Brower
Dr. Audrey K. Brown
Dr. Richard Burger
Dr. John J. Burns
Dr. Michael Burt
Dr. Vincent P. Butler
Dr. Joel N. Buxbaum
Dr. Robert Cancro
Dr. Paul J. Cannon
Dr. Peter W. Carmel
Dr. Arturo Casadevall
Dr. J. Casals-Ariet
Dr. Veronica M. Catanese
Dr. Daniel F. Catanzaro
Dr. Raju S.K. Chaganti
Dr. Moses Victor Chao
Dr. Herbert S. Chase
Dr. Tie Chen
Dr. Leonard Chess
Dr. David Chi
Dr. Raul Chiesa
Dr. Francis P. Chinard*
Dr. Yong Sung Choi
Dr. Purnell W. Choppin
Dr. Dennis J. Cleri
Dr. William L. Cleveland
Dr. David E. Cobrinik

*Life Member.

DR. GERALD COHEN
DR. RANDOLPH P. COLE
DR. MORTON COLEMAN
DR. BARRY S. COLLER
DR. LAWRENCE A. CONE
DR. JOHN A. CONNOR
DR. STUART D. COOK
DR. ARMAND F. CORTESE
DR. PHILIP D. COTTER
DR. JOHN P. CRAIG
DR. EVA BROWN CRAMER
DR. DAVID CRANDALL
DR. BRUCE N. CRONSTEIN
DR. KATHRYN L. CROSSIN
DR. BRUCE CUNNINGHAM
DR. CHARLOTTE CUNNINGHAM-RUNDLES
DR. TOM CURRAN
DR. SAMUEL J. DANISHEFSKY
DR. JAMES E. DARNELL
DR. EARL W. DAVIE
DR. JEAN DAVIGNON
DR. ROBERT P. DAVIS
DR. VITTORIO DEFENDI
DR. RALPH B. DELL
DR. JOHN R. DENTON
DR. RONALD A. DEPINHO
DR. ROBERT J. DESNICK
DR. DICKSON D. DESPOMMIER
DR. PATRICIA A. DETMERS
DR. SALVATORE DI MAURO
DR. BETTY DIAMOND
DR. ALBERTO DIDONATO
DR. ETHAN DMITROVSKY
DR. PHILIP J. DORMAN
DR. R. GORDON DOUGLAS
DR. STEVEN D. DOUGLAS
DR. PETER CLOWES DOWLING
DR. ARNOLD DRAPKIN
DR. PAUL DREIZEN
DR. MICHAEL J. DROLLER
DR. LEWIS M. DRUSIN
DR. RONALD E. DRUSIN
DR. A.E. DUMONT
DR. LAUREL A. ECKHARDT
DR. GERALD M. EDELMAN
DR. NORMAN EDELMAN
DR. RICHARD L. EDELSON
DR. MARK L. EDWARDS
DR. SHIMON EFRAT
DR. HERMAN N. EISEN
DR. ROBERT P. EISINGER
DR. NATHAN AMES ELLIS
DR. SCOTT W. EMMONS
DR. DIANE ESPOSITO
DR. SOLOMON ESTREN
DR. RONALD B. FAANES
DR. THOMAS J. FAHEY
DR. SAUL J. FARBER*
DR. MEHDI FARHANGI
DR. DONALD FISCHMAN
DR. PAUL B. FISHER
DR. PATRICK FITZGERALD
DR. ARTHUR C. FOX
DR. MANFRED FRASCH
DR. BLAIR A. FRASER
DR. LEONARD PAUL FREEDMAN
DR. MICHAEL L. FREEDMAN
DR. ALAN H. FRIEDMAN
DR. ELI A. FRIEDMAN
DR. RONALD FRIEDMAN
DR. WILHELM R. FRISELL
DR. HENRY FURNEAUX
DR. YASUHIRO FURUICHI
DR. W. EINAR GALL
DR. FREDERICK GATES
DR. DONALD A. GERBER
DR. JAMES GERMAN
DR. MARVIN GERSHENGORN
DR. MICHAEL D. GERSHON
DR. MENARD M. GERTLER*
DR. ALLAN GIBOFSKY
DR. HARRIET S. GILBERT
DR. CHARLES GILVARG
DR. HENRY GINSBERG
DR. LEONARD GLASS

*Life Member.

Dr. David L. Globus
Dr. Henry P. Godfrey
Dr. G. Nigel Godson
Dr. Allen M. Gold
Dr. Jonathan W.M. Gold
Dr. Leslie I. Gold
Dr. Marcia B. Goldberg
Dr. David W. Golde
Dr. Lewis Goldfrank
Dr. Laura T. Goldsmith
Dr. David A. Goldstein
Dr. Gideon Goldstein
Dr. Robert Goldstein
Dr. Robert A. Good*
Dr. Emil C. Gotschlich
Dr. Eugene L. Gottfried
Dr. Dicran Goulian
Dr. R.F. Grady
Dr. Lester Grant
Dr. Olga Greengard
Dr. Ezra M. Greenspan
Dr. Anthony J. Grieco
Dr. Randall B. Griepp
Dr. Steven S. Gross
Dr. Melvin M. Grumbach
Dr. Joseph J. Guarneri
Dr. Peter M. Guida
Dr. Guido Guidotti
Dr. Subhash Gulati
Dr. Roy Gulick
Dr. Barry M. Gumbiner
Dr. Sidney Gutstein
Dr. Jack W.C. Hagstrom
Dr. David P. Hajjar
Dr. Katherine A. Hajjar
Dr. Leonard D. Hamilton
Dr. Ulrich Hammerling
Dr. Hideasburo Hanafusa
Dr. Ken Harewood
Dr. Peter C. Harpel
Dr. Michael B. Harris
Dr. Franz-Ulrich Hartl
Dr. George A. Hashim
Dr. Sami Hashim
Dr. Victor Hatcher
Dr. A. Daniel Hauser
Dr. Richard Hawkins
Dr. John Healey
Dr. Samuel Hellman
Dr. Walter L. Henley
Dr. Jules Hirsch
Dr. Kurt Hirschhorn
Dr. Rochelle Hirschhorn
Dr. John L. Ho
Dr. David S. Hodes
Dr. Raymond F. Holden
Dr. Robert S. Holzman
Dr. Marshall S. Horwitz
Dr. Susan Horwitz
Dr. S. Steven Hotta
Dr. Ling-Ling Hsieh
Dr. Howard H. T. Hsu
Dr. Jerard Hurwitz
Dr. Thomas Hutteroth
Dr. Tan Ahmet Ince
Dr. Laura Inselman
Dr. Henry D. Isenberg
Dr. Luis M. Isola
Dr. Fuyuki Iwasa
Dr. Jerry C. Jacobs
Dr. William R. Jacobs
Dr. Eric A. Jaffe
Dr. Herbert Jaffe*
Dr. Alfonso H. Janoski
Dr. Charles I. Jarowski
Dr. Maria Jasin
Dr. Jamshid Javid
Dr. S. Michael Jazwinski
Dr. Edward M. Johnson
Dr. Warren D. Johnson, Jr.
Dr. Richard Jove
Dr. David B. Kaback
Dr. Ronald Kaback
Dr. Lawrence J. Kagan
Dr. Martin L. Kahn
Dr. Melvin Kahn

*Life Member.

Dr. Mikio Kamiyama
Dr. Yoshinobu Kanno
Dr. Gilla Kaplan
Dr. Arthur Karanas
Dr. Arthur Karlin
Dr. Michael Katz
Dr. Herbert J. Kayden
Dr. Gordon I. Kaye
Dr. Patrick Kelly
Dr. Richard H. Kessin
Dr. Shahid M.M. Khan
Dr. Yoon Berm Kim
Dr. Tom Kindt
Dr. Donald West King
Dr. David G. Klapper
Dr. Jack Klatell
Dr. Janis V. Klavins
Dr. Albrecht K. Kleinschmidt
Dr. Joseph A. Kochen
Dr. Samuel S. Koide
Dr. Kiyomi Koizumi
Dr. Richard Neal Kolesnick
Dr. Edwin H. Kolodny
Dr. Thomas R. Kozel
Dr. Philip J. Kozinn
Dr. Rosemary Kraemer
Dr. Laurence Krakoff
Dr. Richard M. Krause
Dr. Alfred N. Krauss
Dr. Robert Krauss
Dr. Gert Kreibich
Dr. Terry Ann Krulwich
Dr. Raju S. Kucherlapati
Dr. Friedrich Kueppers
Dr. Sherman Kupfer
Dr. Marvin Kuschner
Dr. Robert G. Lahita
Dr. Frank R. Landsberger
Dr. Daniel Lane
Dr. Etienne Y. Lasfarques
Dr. Leroy S. Lavine
Dr. Paul B. Lazarow
Dr. Mathew Lee
Dr. Stanley L. Lee*
Dr. Albert M. Lefkovits
Dr. Thomas J.A. Lehman
Dr. Ruth Lehmann
Dr. Neal S. LeLeiko
Dr. Stephen H. Leslie*
Dr. Gerson T. Lesser
Dr. Richard D. Levere
Dr. Richard I. Levin
Dr. Robert A. Levine
Dr. Martin R. Liebowitz
Dr. Stanley L. Lin
Dr. Hsiou-Chi Liou
Dr. George Lipkin
Dr. Martin Lipkin
Dr. Philip Livingston
Dr. John N. Loeb
Dr. Irving M. London*
Dr. Jerome Lowenstein
Dr. David R. Luck
Dr. Lucio Luzzatto
Dr. Daniel Macken
Dr. Richard J. Mahler
Dr. Umadas Maitra
Dr. Mart Mannik
Dr. Aaron J. Marcus
Dr. Philip I. Marcus
Dr. Robert F. Margolskee
Dr. Catherine T. Marino
Dr. Andrew Marks
Dr. Paul A. Marks
Dr. Joan Massague
Dr. H. Robert Masure
Dr. Kenneth S. McCarty
Dr. Maclyn McCarty*
Dr. Paul R. McHugh
Dr. James J. McSharry
Dr. Robert McVie
Dr. Robert B. Mellins
Dr. Donna Mildvan
Dr. Frederick Miller
Dr. C. Richard Minick
Dr. Peter Model

*Life Member.

Dr. Richard C. Mohs
Dr. Malcolm A.S. Moore
Dr. Takashi Morimoto
Dr. Akira Morishima
Dr. Thomas Morris
Dr. John Henry Morrison
Dr. Richard W. Moyer
Dr. Hans J. Mueller-Eberhard
Dr. William A. Muller
Dr. W.P. Laird Myers*
Dr. Ralph L. Nachman
Dr. Gabriel G. Nahas
Dr. Tatsuji Namba
Dr. Morton Nathanson*
Dr. Stanley G. Nathenson
Dr. Brian A. Naughton
Dr. James V. Neel
Dr. Carol Shaw Newlon
Dr. Hanh T. Nguyen
Dr. Warren W. Nichols
Dr. Lee Ann Niswander
Dr. Robert Nolan
Dr. Neil J. Nusbaum
Dr. Ruth Nussenzweig
Dr. Victor Nussenzweig
Dr. Herbert F. Oettgen
Dr. Michiko Okamoto
Dr. Allen I. Oliff
Dr. Peter D. Orahovats
Dr. Zoltan Ovary
Dr. Norbert I.A. Overweg
Dr. Harvey L. Ozer
Dr. Stephen A. Paget
Dr. Elisabeth Paietta
Dr. George E. Palade*
Dr. Peter Palese
Dr. George D. Pappas
Dr. Jean Papps
Dr. Wade P. Parks
Dr. Pedro Pasik
Dr. Mark W. Pasmantier
Dr. Gavril W. Pasternak
Dr. Dinshaw J. Patel
Dr. Pierluigi Patriarca
Dr. Christian C. Patrick
Dr. Nikola P. Pavletich
Dr. Elinor I. B. Peerschke
Dr. Robert G. Pergolizzi
Dr. Richard G. Pestell
Dr. Karl H. Pfenninger
Dr. Mark R. Philips
Dr. Lennart Philipson*
Dr. Julia M. Phillips-Quagliata
Dr. Xavier Pi-Sunyer
Dr. Matthew Pincus
Dr. Steven Maurice Podos
Dr. Beatriz G.T. Pogo
Dr. Jeffrey W. Pollard
Dr. Melissa Pope
Dr. Edwin A. Popenoe*
Dr. Jerome G. Porush
Dr. Jerome B. Posner
Dr. Marshall P. Primack
Dr. John F. Prudden
Dr. Dominick P. Purpura
Dr. Bertha Rader
Dr. Francesco Ramirez
Dr. Jacob Rand
Dr. Helen M. Ranney*
Dr. Jeffrey V. Ravetch
Dr. Lawrence W. Raymond
Dr. George G. Reader
Dr. George E. Reed
Dr. George N. Reeke
Dr. Gabrielle H. Reem
Dr. Westley Reeves
Dr. Joan Reibman
Dr. Marcus M. Reidenberg
Dr. Ronald F. Rieder
Dr. Richard A. Rifkind
Dr. Robert R. Riggio
Dr. Marcos Rivelis
Dr. James L. Roberts
Dr. Hugh D. Robertson
Dr. Alan G. Robinson
Dr. Enrique Rodriguez-Boulan

*Life Member.

DR. ROBERT G. ROEDER
DR. CHARLES E. ROGLER
DR. MARCOS ROJKIND
DR. WILLIAM N. ROM
DR. SHELDON P. ROTHENBERG
DR. JAMES E. ROTHMAN
DR. LEWIS P. ROWLAND
DR. ARYE RUBINSTEIN
DR. MARJORIE RUSSEL
DR. DAVID SABATINI
DR. DAVID B. SACHAR
DR. LETTY G. M. SALENTIJN
DR. MILTON R. J. SALTON
DR. STANLEY SAMUELS
DR. JOHN I. SANDSON
DR. REGINA M. SANTELLA
DR. BIRGIT SATIR
DR. PETER SATIR
DR. BRIJ B. SAXENA
DR. MATTHEW D. SCHARFF
DR. DAVID A. SCHEINBERG
DR. BARBARA M. SCHER
DR. LAWRENCE SCHERR
DR. ALEXANDER SCHIER
DR. PETER B. SCHIFF
DR. GERALD SCHIFFMAN
DR. EDWARD B. SCHLESINGER*
DR. VERN L. SCHRAMM
DR. ULRICH K. SCHUBART
DR. RISE SCHWAB
DR. MORTON K. SCHWARTZ
DR. RICHARD H. SCHWARTZ
DR. MORTON SCHWIMMER
DR. KATHLEEN W. SCOTTO
DR. SAMUEL SEIFTER*
DR. STEPHEN J. SELIGMAN
DR. BRIDGET SHAFIT-ZAGARDO
DR. BRENDA MAE SHANK
DR. ROBERT E. SHANK
DR. WARREN B. SHAPIRO
DR. MICHAEL SHELANSKI
DR. DENNIS SHIELDS
DR. JOYCE E. SHRIVER
DR. BERNARD J. SICURANZA
DR. M.A.Q. SIDDIQUI
DR. GEORGE J. SIEGEL
DR. PHILIP SILVERMAN
DR. EMANUEL SILVERSTEIN
DR. ROY L. SILVERSTEIN
DR. SAUL J. SILVERSTEIN
DR. MICHAEL S. SIMBERKOFF
DR. ERIC J. SIMON
DR. ANNELIESE L. SITARZ
DR. FRANK R. SMITH
DR. JAMES P. SMITH
DR. ROSEMARY SOAVE
DR. STEFAN SOMLO
DR. EDMUND H. SONNENBLICK
DR. A. SPECTOR
DR. FRANK C. SPENCER
DR. DEMPSEY SPRINGFIELD
DR. P.R. SRINIVASAN
DR. E. RICHARD STANLEY
DR. PAMELA STANLEY
DR. NEAL H. STEIGBIGEL
DR. CHARLES R. STEINMAN
DR. RALPH M. STEINMAN
DR. KURT H. STENZEL
DR. PETER E. STOKES
DR. WILLIAM T. STUBENBORD
DR. FREDERICK JOSEPH SUCHY
DR. STEPHEN K. SULLIVAN
DR. BARNET M. SULTZER
DR. MANIKKAM SUTHANTHIRAN
DR. YASUNOSUKE SUZUKI
DR. ROY C. SWINGLE
DR. MARGUERITE P. SYKES
DR. WILLIAM TALBOT
DR. J.R. TATA
DR. ALVIN TEIRSTEIN
DR. NITIN T. TELANG
DR. G. JEANETTE THORBECKE
DR. DAVID A. TICE
DR. JIR SHIONG TSAI
DR. GERARD M. TURINO
DR. KOHJI UCHIZONO

*Life Member.

Dr. Hiroshi Ueno
Dr. Jay C. Unkeless
Dr. Fred T. Valentine
Dr. Mary A. Vogler
Dr. John A. Wagner
Dr. Lila A. Wallis*
Dr. Jen C. Wang
Dr. Jonathan R. Warner
Dr. Samuel Waxman
Dr. Josiah F. Wedgwood
Dr. Richard Weil
Dr. Jeffrey Weinreb
Dr. Harel Weinstein
Dr. I. B. Weinstein
Dr. Gerson Weiss
Dr. Daniel Wellner
Dr. Martin Wiedmann
Dr. Peter H. Wiernik
Dr. Torsten N. Wiesel
Dr. Murray Wittner
Dr. David Wolf
Dr. Savio L.C. Woo
Dr. Peter I. Yi
Dr. Michael W. Young
Dr. Ralph Zalusky
Dr. Liang Zhu
Dr. Norton D. Zinder
Dr. Dorothea Zucker-Franklin
Dr. Arturo Zychlinsky

*Life Member.